Handbook of ALGORITHMS and DATA STRUCTURES

INTERNATIONAL COMPUTER SCIENCE SERIES

Handbook of ALGORITHMS and DATA STRUCTURES

G. H. Gonnet

ADDISON-WESLEY PUBLISHING COMPANY
London · Reading, Massachusetts · Menlo Park, California · Amsterdam
Don Mills, Ontario · Manila · Singapore · Sydney · Tokyo

Set by the author on an APS-5 photosetter using output generated by troff, the UNIX text-processing system (UNIX is a trademark of Bell Laboratories)

Cover illustration produced by John Vince using PICASO, the computer graphics system at Middlesex Polytechnic

Printed in Finland by OTAVA. Member of Finnprint

OTABIND
PAT. PEND.

British Library Cataloguing in Publication Data

Gonnet, G. H.
Handbook of algorithms and data structures.
1. Electronic digital computers—Programming
2. Algorithms
I. Title
001.6′42 QA76.6

ISBN 0-201-14218-X

Library of Congress Cataloging in Publication Data

Gonnet, G. H. (Gaston H.)
Handbook of algorithms and data structures.
(International computer science series)
Bibliography: p.
Includes index.
1. Electronic digital computers—Programming.
2. Algorithms. 3. Data structures (Computer science)
I. Title. II. Series.
QA76.6.G636 1984 001.64′2 83-27301

ISBN 0-201-14218-X

ABCDEF8987654

To my boys: Miguel, Pedro, Julio and Ignacio
and my girls: Araina and Marte

Preface

Computer Science has been, throughout its evolution, more an art than a science. My favourite example which illustrates this point is to compare a major software project (like the writing of a compiler) with any other major project (like the construction of the CN tower in Toronto). It would be absolutely unthinkable to let the tower fall down a few times while its design was being debugged: even worse would be to open it to the public before discovering some other fatal flaw. Yet this mode of operation is being used everyday by almost everybody in software production.

Presently it is very difficult to "stand on your predecessor's shoulders", most of the time we stand on our predecessor's toes, at best. This handbook was written with the intention of making available to the computer scientist, instructor or programmer the wealth of information which the field has generated in the last 20 years.

Most of the results are extracted from the given references. In some cases the author has completed or generalized some of these results. Accuracy is certainly one of our goals, and consequently the author will cheerfully pay $2.00 for each first report of any type of error appearing in this handbook.

Many people helped me directly or indirectly to complete this project. Firstly I owe my family hundreds of hours of attention. All my students and colleagues had some impact. In particular I would like to thank M. C. Momard, N. Ziviani, J. I. Munro, P. A. Lanson, D. Rotem and D. Wood. Very special thanks go to F. W. Tompa who is also the coauthor of chapter 2. The source material for this chapter appears in a joint paper in the November 1983 *Communications of the ACM*.

Montevideo
December 1983

G. H. Gonnet

Contents

Chapter 1: INTRODUCTION

This handbook is intended to contain most of the information available on algorithms and their data structures; thus it is designed to serve a wide spectrum of users, from the programmer who wants to code efficiently to the student or researcher who needs quick information.

The main emphasis is placed on algorithms. For these we present their description, code in one or more languages, theoretical results and extensive lists of references.

1.1 Structure of the chapters

The handbook is organized by topics. The second chapter offers a formalization of the description of algorithms and data structures, chapters 3 to 6 discuss searching, sorting, selection and arithmetic algorithms respectively. Appendix I describes some probability distributions encountered in data processing; appendix II contains a collection of asymptotic formulas related to the analysis of algorithms; appendix III contains the main list of references and appendix IV contains alternate code for some algorithms.

The chapters describing algorithms are divided in sections and subsections as needed. Each algorithm is described in its own sub-section, and all have roughly the same format, though we may make slight deviations or omissions when information is unavailable or trivial. The general format includes:

(1) Definition and explanation of the algorithm and its classification according to the basic operations described in chapter 2.

(2) Theoretical results on the algorithm's complexity. We are mainly interested in measurements which indicate an algorithm's running time and its space requirements. Useful quantities to measure for this information include the number of comparisons, data accesses, assignments, or exchanges an algorithm might make. When looking at space requirements, we might consider the number of words, records, or pointers involved in an implementation. Time complexity covers a much broader range of measurements. For example, in our examination of searching algorithms, we might be able to attach

meaningful interpretations to most of the combinations of the

$$\left\{\begin{array}{c}\text{average}\\ \text{variance}\\ \text{minimum}\\ \text{worst}-\text{case}\\ \text{average w.c.}\end{array}\right\} \text{number of} \left\{\begin{array}{c}\text{comparisons}\\ \text{accesses}\\ \text{assignments}\\ \text{exchanges}\\ \text{function calls}\end{array}\right\} \text{when we} \left\{\begin{array}{c}\text{query}\\ \text{add a record into}\\ \text{delete a record from}\\ \text{modify a record of}\\ \text{reorganize}\\ \text{build}\\ \text{read sequentially}\end{array}\right\}$$

the structure. Other theoretical results may also be presented, such as enumerations, generating functions, or behaviour of the algorithm when the data elements are distributed according to special distributions.

(3) The Algorithm. We have selected Pascal and C to describe the algorithms. Algorithms that may be used in practice are described in one or both of these languages. For algorithms which are only of theoretical interest, we do not provide their code. Algorithms which are coded both in Pascal and in C will have one code with the algorithm and the other in appendix IV.

(4) Recommendations. Following the algorithm description we give several hints and tips on how to use the algorithm. We point out pitfalls to avoid in coding, suggest when to use the algorithm and when not to, tell when to expect best and worst performances, and provide a variety of other comments.

(5) Tables. Whenever possible, we present tables which show exact values of complexity measures in selected cases. These are intended to give a feeling for how the algorithm behaves. When precise theoretical results are not available we give simulation results, generally in the form $xxx \pm yy$ where the value yy is chosen so that the resulting interval has a confidence level of 95%. In other words, the actual value of the complexity measure falls out of the given interval only once every twenty simulations.

(6) Differences between internal and external storage. Some algorithms may perform better for internal storage than external, or vice versa. When this is true, we will give recommendations for applications in the two different cases. Since most of our analysis up to this point implicitly assumes that internal memory is used, in this section we will look more closely at the external case (if appropriate). We analyze the algorithm's behaviour when working with external storage, and discuss any significant practical considerations in using the algorithm externally.

(7) With the description of each algorithm we include a list of relevant references. General references, surveys, or tutorials are collected at the end of chapters or sections. The second appendix contains an alphabetical list of all references with cross-references to the relevant algorithms.

1.2 Naming of variables

The naming of variables throughout this handbook is a compromise between uniformity of notation and accepted terminology in the specific areas.

Except for very few exceptions, explicitly noted, we use:

n for the number of objects or elements or components in a structure;

m for the size of a structure;

b for bucket sizes, or maximum number of elements in a physical block;

d for the digital cardinality or size of the alphabet.

The complexity measures are also named uniformly throughout the handbook. Complexity measures are named X_n^Z and should be read as "the number of X's performed while doing Z onto a structure of size n".

Typical values for X are:

A: accesses, probes or node inspections;

C: comparisons or node inspections;

E: external accesses;

h: height of a recursive structure (typically a tree);

I: iterations (or number of function calls);

L: length (of path or longest probe sequence);

M: moves or assignments (usually related to record or key movements).

Typical values for Z are:

null (no superscript): successful search (or default operation, when there is only one possibility);

C: construction (building) of structure;

D: deletion of an element;

E: extraction of an element (mostly for priority queues);

I: insertion of a new element;

M: merging of structures;

Opt: optimal construction or optimal structure (the operation is usually implicit);

MM: minimax, or minimum number of X's in the worst case: this is usually used to give upper and lower bounds on the complexity of a problem.

Note that X_n^I means number of operations done to insert an element into a structure of size n or to insert the $n+1^{st}$ element.

Although these measures are random variables (as these depend on the particular structure on which they are measured), we will make exceptions for C_n and C_n' which most of the literature considers to be expected values.

1.3 Probabilities

The probability of a given event is denoted by $Pr\{event\}$. Random variables follow the convention described in the preceding section. The expected value of a random variable X is written $E[X]$ and its variance is $\sigma^2(X)$. In particular, for discrete variables X

$$E[X] = \sum_i i\, Pr\{X=i\}$$

$$\sigma^2(X) \;=\; \sum_i i^2 Pr\{X=i\} - E[X]^2 \;=\; E[X^2] - E[X]^2$$

We will always make explicit the probability universe on which expected values are computed. This is ambiguous in some cases, and is a ubiquitous problem with expected values.

To illustrate the problem without trying to confuse the reader, suppose that we fill a hashing table with keys and then we want to know about the average number of accesses to retrieve one of the keys. We have two potential probability universes: the key selected for retrieval (the one inserted first, the one inserted second, etc.) and the actual values of the keys, or their probing sequence. We can compute expected values with respect to the first, the second, or both universes. In simpler terms, we can find the expected value of any key for a given file, or the expected value of a given key for any file, or the expected value of any key for any file.

Unless otherwise stated, (i) the distribution of our elements is always random independent uniform $U(0,1)$; (ii) the selection of a given element is uniform discrete between all possible elements; (iii) expected values which relate to multiple universes are computed with respect to all universes. In terms of the above example, we will compute expected values with respect to randomly selected variables drawn from a uniform $U(0,1)$ distribution.

1.4 Asymptotic notation

Most of the complexity measures in this handbook are asymptotic in the size of the problem. The asymptotic notation we will use is fairly standard and is given below:

$$f(n) \;=\; O(g(n))$$

implies that there exists k and n_0 such that $|f(n)| < kg(n)$ for $n > n_0$

$$f(n) \;=\; o(g(n)) \;\longrightarrow\; \lim_{n\to\infty} \frac{f(n)}{g(n)} = 0$$

$$f(n) \;=\; \Theta(g(n))$$

implies that there exists k_1, k_2, $(k_1 \times k_2 > 0)$ and n_0 such that $k_1 g(n) < f(n) < k_2 g(n)$ for $n > n_0$, or equivalently that $f(n) = O(f(n))$ and $g(n) = O(f(n))$.

$$f(n) \;=\; \Omega(g(n)) \;\longrightarrow\; g(n) \;=\; O(f(n))$$

$$f(n) \;=\; \omega(g(n)) \;\longrightarrow\; g(n) \;=\; o(f(n))$$

$$f(n) \;\approx\; g(n) \;\longrightarrow\; f(n) - g(n) \;=\; o(g(n))$$

We will freely use arithmetic operations with the order notation, e.g.

$$f(n) \;=\; h(n) + O(g(n))$$

means

$$f(n) - h(n) = O(g(n))$$

Whenever we write $f(n) = O(g(n))$ it is with the understanding that we know of no better asymptotic bound, i.e. we know of no $h(n) = o(g(n))$ such that $f(n) = O(h(n))$.

1.5 About the programming languages

We use two languages to code our algorithms: Pascal and C. After writing many algorithms we still find situations for which these languages do not present a very "clean" or understandable code. Therefore, whenever possible, we use the language which presents the shortest and most readable code. We intentionally allow our Pascal and C style of coding to resemble each other.

A minimal number of Pascal programmes contain **goto** statements. These statements were used in place of the equivalent C statements **return** and **break**, and are correspondingly so commented. Indeed we view their absence from Pascal as a shortcoming of the language. Another irritant in coding some algorithms in Pascal is the lack of order in the evaluation of logical expressions. This is unfortunate since such a feature makes algorithms easier to understand. The typical stumbling block is

```
while ( p ≠ nil ) and ( key ≠ p↑.key ) do ....
```

Such a statement works in C if we use the sequential and operator (&&), but for Pascal we have to use instead:

```
while p ≠ nil do begin
        if key = p↑.k then goto 999 {*** break ***};
        ....
999:
```

Other minor objections are: the inability to compute addresses of non-heap objects in Pascal (which makes treatment of lists more difficult); the lack of a **with** statement in C and the lack of **var** parameters in C. (Although this is technically possible to overcome, it obscures the algorithms.)

Our Pascal code conforms, as fully as possible, to the language described in *Pascal User Manual and Report* by K. Jensen and N. Wirth. The C code conforms to the language described in *The C Programming Language* by B.W. Kernighan and D.M. Ritchie.

1.6 On the code for the algorithms

Except for very few algorithms which are obviously written in pseudo-code, the algorithms in this handbook were run and tested under two different compilers. Actually the same text which is printed is used for compiling, for testing, for running simulations and for obtaining timings. This was done in an attempt to eliminate (or at least drastically reduce!) errors.

Each family of algorithms has a "tester set" which not only checks for correct behaviour of the algorithm, but also checks proper handling of limiting conditions (will a sorting routine sort a null file? one with one element? one with all equal keys? etc.).

In most cases the algorithms are described as a function or a procedure or a small set of functions or procedures. In a few cases, for very simple algorithms, the code is described as in-line code, which could be encapsulated in a procedure or could be inserted into some other piece of code.

Some algorithms, most notably the searching algorithms, are building blocks or components of other algorithms or programmes. Some standard actions should not be specified for the algorithm itself, but rather will be specified once that the algorithm is "composed" with other parts (chapter 2 defines composition in more detail). A typical example of a standard action is an error condition. The algorithms coded for this handbook use always the same names for these standard actions.

Error detection of an unexpected condition during execution. Whenever *Error* is encountered it can be substituted by any block of statements. For example our testers print an appropriate message.

found(record) function call that is executed upon completion of a successful search. Its argument is a record or a pointer to a record which contains the searched key.

notfound(key) function called upon an unsuccessful search. Its argument is the key which was not found.

A special effort has been made to avoid duplication of these standard actions for identical conditions. This makes it easier to substitute blocks of code for these.

1.7 Complexity measures and real timings

For some families of algorithms we include a comparison of real timings. These timings are to be interpreted with caution as they reflect only one sample point in the many dimensions of hardwares, compilers, operating systems, etc. Yet we have equally powerful reasons to present at least one set of real complexities.

The main reasons for including real timing comparisons are:

(i) these take into account the actual cost of operations,

(ii) these take into account hidden costs, like storage allocation, indexing, etc.

The main objections, or the factors which may invalidate these real timing tables, are:

(i) the results are compiler dependent: although the same compiler is used for each language, a compiler may favour some construct over others;

(ii) the results are hardware dependent,

(iii) in some cases, when large amounts of memory are used, the timings may be load dependent.

The timings were done on a VAX 11/780 running the Berkeley Unix 4.1 operating system. Both C and Pascal compilers were run with the optimizer, or object code improver, to obtain the best implementation for the algorithms.

There were no attempts made to compare timings across languages. All the timing results are computed relative to the fastest algorithm. To avoid the incidence of start up costs, loading, etc. the tests were run on problems of significant size. Under these circumstances, some $O(n^2)$ algorithms appear to perform very poorly.

Chapter 2: BASIC CONCEPTS

2.1 Data structure description

The formal description of data structure implementations is similar to the formal description of programming languages. In defining a programming language, one typically begins by presenting a syntax for valid programmes in the form of a grammar and then sets further validity restrictions (e.g., usage rules for symbolic names) which give constraints that are not captured by the grammar. Similarly, a valid data structure implementation will be one that satisfies a syntactic grammar and also obeys certain constraints. For example, for a particular data structure to be a valid weight balanced binary tree, it must satisfy the grammatical rules for binary trees and it must also satisfy a specific balancing constraint.

2.1.1 Grammar for data objects

A sequence of real numbers can be defined by the BNF production

<S> ::= [real , <S>] | nil

Thus a sequence of reals can have the form nil, [real,nil], [real,[real,nil]], and so on. Similarly, sequences of integers, characters, strings, boolean constants, etc. could be defined. However, this would result in a bulky collection of production rules which are all very much alike. One might first try to eliminate this repetitiveness by defining

<S> ::= [<D> , <S>] | nil

where <D> is given as the list of data types

<D> ::= real | int | bool | string | char

However, this pair of productions generates unwanted sequences such as

[real,[int,nil]]

as well as the homogeneous sequences desired.

To overcome this problem, the syntax of a data object class can be defined using a W-grammar (also called a two-level or van Wijngaarden grammar). Actually the full capabilities of W-grammars will not be utilized; rather the syntax will be defined using the equivalent of standard BNF productions together with the uniform replacement rule as described below.

A W-grammar generates a language in two steps (levels). In the first step, a collection of generalized rules is used to create more specific production rules. In the second step, the production rules generated in the first step are used to define the actual data structures.

First, the problem of listing repetitive production rules is solved by starting out with generalized rule-forms known as *hyperrules*, rather than the rules themselves. The generalized form of a sequence S is given by the hyperrule

s − D : [D, s − D] ; nil

The set of possible substitutions for **D** are now defined in a *metaproduction*, as distinguished from a conventional BNF-type production. For example, if **D** is given as

D :: real; int; bool; string; char; · · ·

a sequence of real numbers is defined in two steps as follows. The first step consists of choosing a value to substitute for **D** from the list of possibilities given by the appropriate metaproduction; in this instance, **D** → **real.** Next invoke the uniform replacement rule to substitute the string **real** for **D** everywhere it appears in the hyperrule that defines **s − D**. This substitution gives

s − real : [real , s − real] ; nil

Thus the joint use of the metaproduction and the hyperrule generates an ordinary BNF-like production defining real sequences. The same two statements can generate a production rule for sequences of any other valid data type (integer, character, etc.).

Figures 2.1 and 2.2 contain a W-grammar which will generate many conventional data objects. As further examples of the use of this grammar, consider the generation of a binary tree of real numbers. With **D→real** and **LEAF→nil**, HR[3] generates the production rule

bt − real − nil : [real , bt − real − nil , bt − real − nil] ; nil

Since **bt − real − nil** is one of the legitimate values for **D** according to M[1] let **D→bt − real − nil** from which HR[1] indicates that such a binary tree is a legitimate data structure.

Secondly consider the specification for a hash table to be used with direct chaining. The production

s − (string,int) : [(string,int) , s − (string,int)] ; nil

and M[1] yield

$$\mathbf{D} \rightarrow \{\mathbf{s-(string,int)}\}_0^{96}$$

Thus HR[1] will yield a production for an array of sequences of string/integer pairs usable, for example, to record NAME/AGE entries using hashing.

Finally consider a production rule for structures to contain B-trees of strings using HR[4] and the appropriate metaproductions to yield

$$\mathbf{mt-10-string-nil : [int, \{string\}_1^{10}, \{mt-10-string-nil\}_0^{10}] \; ; \; nil}$$

Metaproductions:

M[1]	**D** ::	**real;int;bool;string;char;...;**	# atomic data types
		$\{$ **D** $\}_N^N$;	# array
		REC ; (REC) ;	# record
		[D] ;	# reference
		s−D;	# sequence
		gt−D−LEAF;	# general tree
		DICT;	# dictionary structures
		... .	# other structure classes E.g. graphs, sets, priority queues.
M[2]	**DICT** ::	$\{$**KEY**$\}_N^N$; **s−KEY;**	# sequential search
		bt−KEY−LEAF;	# binary tree
		mt−N−KEY−LEAF;	# multi-way tree
		tr−N−KEY.	# trie
M[3]	**REC** ::	**D; D, REC.**	# record definition
M[4]	**LEAF** ::	**nil; D.**	
M[5]	**N** ::	**DIGIT; DIGIT N.**	
M[6]	**DIGIT** ::	**0;1;2;3;4;5;6;7;8;9.**	
M[7]	**KEY** ::	**real;int;string;char;(KEY,REC).**	# search key

Figure 2.1: Metaproductions for data objects

Hyperrules:

HR[1]	**data structure :**	**D.**
HR[2]	**s−D :**	**[D,s−D] ; nil.**
HR[3]	**bt−D−LEAF :**	**[D,bt−D−LEAF,bt−D−LEAF] ; LEAF.**
HR[4]	**mt−N−D−LEAF :**	**[int,**$\{$**D**$\}_1^N$**,**$\{$**mt−N−D−LEAF**$\}_0^N$**] ; LEAF.**
HR[5]	**gt−D−LEAF :**	**[D,s−gt−D−LEAF] ; LEAF.**
HR[6]	**tr−N−D:**	**[** $\{$ **tr−N−D** $\}_1^N$ **] ; [D] ; nil.**

Figure 2.2: Hyperrules for data objects

In this multitree, each node contains ten keys and has eleven descendants. Certain restrictions on B-trees, however, are not included in this description (that the number of actual keys is to be stored in the **int** field in each node, that this number must be between 5 and 10, that the actual keys will be stored contiguously in the keys-array starting at position 1, etc.); these will instead be defined as constraints (see below).

The grammar rules that we are using are inherently ambiguous. This is not inconvenient; as a matter of fact it is even desirable. For example, consider

$$\mathbf{D} \rightarrow \{\mathbf{D}\}_{\mathbf{N}}^{\mathbf{N}} \rightarrow \{\mathbf{real}\}_{1}^{10} \tag{i}$$

and

$$\mathbf{D} \rightarrow \mathbf{DICT} \rightarrow \{\mathbf{KEY}\}_{\mathbf{N}}^{\mathbf{N}} \rightarrow \{\mathbf{real}\}_{1}^{10} \tag{ii}$$

Although both derivation trees produce the same object, the second one describes an array used as a sequential implementation of a dictionary structure, while the first may just be a collection of real numbers. In other words, the derivation tree used to produce the data objects contains important semantic information and should not be ignored.

2.1.2 Constraints for data objects

Certain syntactic characteristics of data objects are difficult or cumbersome to define using formal grammars. A semantic rule or constraint may be regarded as a boolean function on data objects ($S:D \rightarrow$bool) that indicates which are valid and which are not. Objects that are valid instances of a data structure implementation are those in the intersection of the set produced by the W-grammars and those that satisfy the constraints.

Below are some examples of semantic rules which may be imposed on data structures. As phrased, these constraints are placed on data structures that have been legitimately produced by rules given in the previous section.

2.1.2.1 Sequential order

Many data structures are kept in some fixed order (e.g. the records in a file are often arranged alphabetically or numerically according to some key). Whatever work is done on such a file should not disrupt this order. This definition normally applies to **s−D** and $\{\mathbf{D}\}_{\mathbf{N}}^{\mathbf{N}}$.

2.1.2.2 Uniqueness

Often it is convenient to disallow duplicate values in a structure, for example in representing sets. Other times the property of uniqueness can be used to ensure that records are not referenced several times in a structure (e.g. that a linear chain has no cycles or that every node in a tree has only one parent).

2.1.2.3 Hierarchical order

For all nodes, the value stored at any adjacent node is related to the value at the node according to the type of adjacency. This definition normally applies to **bt−D−LEAF**, **mt−N−D−LEAF** and **gt−D−LEAF**.

Lexicographical trees:

A lexicographical tree is a tree that satisfies the following condition for every node s: if s has n keys ($key_1, key_2, ..., key_n$) stored in it, s must have $n+1$ descendant subtrees $t_0, t_1, \ldots, t_n$. Furthermore, if d_0 is any key in any node of t_0, d_1 any key in any node of t_1, and so on, the inequality $d_0 \leq key_1 \leq d_1 \leq ... \leq key_n \leq d_n$ must hold.

Priority queues:

A priority queue can be any kind of recursive structure in which an order relation has been established between each node and its descendants. One example of such an order relation would be to require that $key_p \leq key_d$, where key_p is any key in a parent node, and key_d is any key in any descendant of that node.

2.1.2.4 Hierarchical balance

Height balance:

Let s be any node of a tree (binary or multi-way). Define $h(s)$ as the height of the subtree rooted in s, i.e. the number of nodes in the tallest branch starting at s. One structural quality that may be required is that the height of a tree along any pair of adjacent branches be approximately the same. More formally, the height balance constraint is $|h(s_1) - h(s_2)| \leq \delta$ where s_1 and s_2 are any two subtrees of any node in the tree, and δ is a constant giving the maximum allowable height difference. In B-trees for example, $\delta = 0$, while in AVL-trees $\delta = 1$.

Weight balance:

For any tree, the weight function $w(s)$ is defined as the number of external nodes (leaves) in the subtree rooted at s. A weight balance condition requires that for any two nodes s_1 and s_2, if they are both subtrees of any other node in the tree, $r \leq w(s_1)/w(s_2) \leq 1/r$ where r is a positive constant less than 1.

2.1.2.5 Optimality

Any condition on a data structure which minimizes a complexity measure (such as the expected number of accesses or the maximum number of comparisons) is an optimality condition. If this minimized measure of complexity is based on a worst-case value, the value is called the *minimax;* when the minimized complexity measure is based on an average value, it is the *minave.*

In summary, the W-grammars are used to define the general shape or pattern of the data objects. Once an object is generated, its validity is checked against the semantic rules or constraints that may apply to it.

References:

[Burton, 79], [Claybrook, 77], [Cremers *et al.*, 78], [O'Dunlaing *et al.*, 82], [Rosenberg *et al.*, 79], [Rosenberg, 75], [Vuillemin, 80].

2.2 Algorithm descriptions

Having defined the objects used to structure data, it is appropriate to describe the algorithms that access them. Furthermore, because data objects are not static, it is equally important to describe data structure manipulation algorithms.

An algorithm computes a function that operates on data structures. More formally, an algorithm describes a map $S \rightarrow R$ or $S \times P \rightarrow R$, where S, P, and R are all data structures; S is called the input structure, P contains parameters (e.g., to specify a query), and R is the result. The two following examples illustrate these concepts:

(1) Quicksort is an algorithm that takes an array and sorts it. Since there are no parameters,

Quicksort: array $\rightarrow$ sorted-array

(2) B-tree insertion is an algorithm that inserts a new record P into a B-tree S, giving a new B-tree as a result. In functional notation,

B-tree-insertion: B-tree $\times$ new-record $\rightarrow$ B-tree

Algorithms compute functions over data structures. As always, different algorithms may compute the same functions; $\sin(2x)$ and $2\sin(x)\cos(x)$ are two expressions that compute the same function. Since equivalent algorithms have different computational requirements however, it is not merely the function computed by the algorithm that is of interest, but also the algorithm itself.

In the following section, we describe a few basic operations informally in order to convey their flavour.

References:
[Bentley *et al.*, 80], [Bentley, 79], [Bentley, 79], [Leeuwen, 80], [Leeuwen, 80], [Mehlhorn *et al.*, 81], [Overmars *et al.*, 81], [Overmars *et al.*, 81], [Overmars *et al.*, 81], [Overmars *et al.*, 81], [Overmars *et al.*, 82], [Overmars, 81], [Remy, 80], [Rosenberg, 81], [Saxe *et al.*, 79].

2.2.1 Basic (or atomic) operations

A primary class of basic operations manipulate atomic values and are used to focus an algorithm's execution on the appropriate part(s) of a composite data object. The most common of these are as follows:

Selector and Constructor

A selector is an operation that allows access to any of the elements corresponding to the right-hand-side of a production rule from the corresponding left-hand-side object. A constructor is an operation that allows us to assemble an element on the left-hand-side of a production given all the corresponding elements on the right. For example, given a $\{\textbf{string}\}_1^5$ and an integer, we can select the ith element, and given two **bt−real−nil** and a **real** we can construct a new **bt−real−nil**.

Replacement: non-scalar $\times$ selector $\times$ value $\rightarrow$ non-scalar

A replacement operator removes us from pure functions by introducing the assignment statements. This operator introduces the possibility of cyclic and shared structures. For example, given a **bt-D-LEAF** we can form a threaded binary tree by replacing the **nil** values in the leaves by (tagged) references back to appropriate nodes in the tree.

Ranking: set of scalars $\times$ scalar $\rightarrow$ integer

This operation is defined on a set of scalars $X_1, X_2, \ldots, X_n$ and uses another scalar X as a parameter. Ranking determines how many of the X_j values are less than or equal to X, thus determining what rank X would have if it were ordered with the other values. More precisely, ranking is finding an integer i such that there is a subset $A \subseteq \{X_1, X_2, \ldots, X_n\}$ for which $|A| = i$ and $X_j \in A$ if and only if $X_j \leq X$. Ranking is used primarily in directing multi-way decisions. For example, in a binary decision, $n=1$, and i is zero if $X<X_1$, one otherwise.

Hashing: value $\times$ range $\rightarrow$ integer
Hashing is an operation which normally makes use of a record key. Rather than using the actual key value however, an algorithm invokes hashing to transform the key into an integer in a prescribed range by means of a hashing function and then uses the generated integer value.

Interpolation: numeric-value $\times$ parameters $\rightarrow$ integer
Similarly to hashing, this operation is typically used on record keys. Interpolation computes an integer value based on the input value, the desired range, the values of the smallest and largest of a set of values, and the probability distribution of the values in the set. Interpolation normally gives the statistical mode of the location of a desired record in a random ordered file, i.e. the most probable location of the record.

Digitization: scalar $\rightarrow$ sequence of scalars
This operation transforms a scalar into a sequence of scalars. Numbering systems that allow the representation of integers as sequences of digits and strings as sequences of characters provide natural methods of digitization.

Testing for equality: value $\times$ value $\rightarrow$ boolean
Rather than relying on multi-way decisions to test two values for equality, a distinct operation is included in the basic set. Given two values of the same type (e.g., two integers, two characters, two strings), this operation determines whether they are equal. Notice that the use of multi-way branching plus equality testing closely matches the behaviour of most processors and programming languages which require two tests for a three-way branch (less than, equal, or greater than).

2.2.2 Building procedures

Building procedures are used to combine basic operations and simple algorithms to produce more complicated ones. In this section, we will define four building procedures.

2.2.2.1 Composition

Composition is the main procedure for producing algorithms from atomic operations. Typically, but not exclusively, the composition of $F_1{:}S \times P \rightarrow R$ and $F_2{:}S \times P \rightarrow R$ can be expressed in a functional notation as $F_2(F_1(S,P_1),P_2)$. A more general and hierarchical description of composition is that the description of F_2 uses F_1 instead of a basic operation.

Although this definition is enough to include all types of composition, there are several common forms of composition that deserve to be identified explicitly.

Divide and conquer
uses a composition involving two algorithms for any problems that are greater than a critical size. The first algorithm splits a problem into (usually two) smaller problems. The composed algorithm is then recursively applied to each non-empty component, using recursion termination (see below) when

appropriate. Finally the second algorithm is used to assemble the components' results into one result. A typical example of divide and conquer is Quicksort (where the termination alternative may use a Linear insertion sort). Diagrammatically:

Divide and conquer

```
solve-problem(A):
    if size(A) ≤ Critical-Size
        then  End-Action
        else  begin
              Split-problem;
              solve-problem(A1);
              solve-problem(A2);
              . . . .
              Assemble-Results
              end;
```

Special cases of divide and conquer, when applied to trees are tree traversals.

Iterative application
operates on an algorithm and a sequence of data structures. The algorithm is iteratively applied using successive elements of the sequence in place of the single element for which it was written. For example, insertion sort iteratively inserts an element into a sorted sequence.

Iterative application

```
solve-problem(S):
    while not empty( S ) do begin
        Apply algorithm to next element of sequence S;
        Advance S
        end;
    End-Action
```

Alternatively, if the sequence is in an array:

Iterative application (arrays)

```
solve-problem(A):
        for i:=1 to size(A) do
                Action on A[i];
        End-Action
```

Tail recursion
is a composition involving one algorithm that specifies the criterion for splitting a problem into (usually two) components and selecting one of them to be solved recursively. A classical example is binary search.

Tail recursion

```
solve-problem(A):
        if size(A) ≤ Critical-Size
                then  End-Action
                else  begin
                      Split and select subproblem i;
                      solve-problem(A_i)
                      end
```

Alternatively, we can unwind the recursion into a while loop:

Tail recursion

```
solve-problem(A):
        while size(A) > Critical-Size do begin
                Split and select subproblem i;
                A := A_i
                end;
        End-Action
```

It should be noted that tail recursion can be viewed as a variant of divide and conquer in which only one of the subproblems is solved recursively. Both *divide and conquer* and *tail recursion* split the original problem into subproblems of the same type. This splitting applies naturally to recursive data structures such as binary trees, multiway trees, general trees, tries, or arrays.

Inversion
is the composition of two search algorithms that are then used to search for sets of records based on values of secondary keys. The first algorithm is used to search for the selected attribute (e.g. find the "inverted list" for the attribute "hair colour" as opposed to "salary range") and the second algorithm is used to search for the set with the corresponding key value (e.g. "blonde" as opposed to "brown"). In general, inversion returns a set of records which may be further processed (for example, using intersection, union, or set difference).

Inverted search

```
inverted-search(S, A, V):
        {*** Search the value V of the attribute A in
                the structure S ***}
        search ( search(S, A), V )
```

The structure S on which the inverted search operates has to reflect these two searching steps. For the generation of S, the following metaproductions should be used:

$$\mathbf{S} \rightarrow \mathbf{D} \rightarrow \mathbf{DICT} \rightarrow \cdots (\mathbf{KEY}^{\mathbf{attr}}, \mathbf{D}^{\mathbf{attr}}) \cdots$$

$$\mathbf{D}^{\mathbf{attr}} \rightarrow \mathbf{DICT} \rightarrow \cdots (\mathbf{KEY}^{\mathbf{value}}, \mathbf{D}^{\mathbf{value}}) \cdots$$

$$\mathbf{D}^{\mathbf{value}} \rightarrow \mathbf{SET} \rightarrow \cdots$$

Digital decomposition
is applied to a problem of size n by attacking preferred-size pieces (for example, pieces of size equal to a power of two). An algorithm is applied to all these pieces to produce the desired result. One typical example is binary decomposition.

Digital decomposition

Solve−problem(A, n):
    {*** n *has a digital decomposition:* $n = n_k\beta_k + ... + n_1\beta_1 + n_0$ ***}
    Partition the problem into subsets

$$A = \bigcup_{i=0}^{k} \bigcup_{j=1}^{n_i} A_i^j;$$

        {*** where size(A_i^j) = β_i ***}
    for i:=0 **to** k **while not** *completed* **do**
        simpler−solve($A_i^1, A_i^2, \ldots, A_i^{n_i}$);

Merge
applies an algorithm and a discarding rule to two or more sequences of data structures ordered on a common key. The algorithm is iteratively applied using successive elements of the sequences in place of the single elements for which it was written. The discarding rule controls the iteration process. For example, set union, intersection, merge sort, and the majority of business applications use merging.

Merge

```
Merge(S_1,S_2,...,S_k):
    while at least one S_i is not empty do
        kmin := minimum value of keys in S_1,...,S_k;
        for i:=1 to k do
            if kmin = head(S_i)
                then t[i] := head(S_i)
                else t[i] := nil;
        processing-rule( t[1], t[2], ..., t[k] );
    End-Action
```

Randomization
is used to improve a procedure or to transform a procedure into a probabilistic algorithm. This is appealing when the underlying procedure may fail, may not terminate, or may have a very bad worst case.

Randomization

```
Solve-problem ( A )
    repeat begin
        randomize( A );
        solve( randomized(A), t(A) units-of-time );
        end    until Solve-Succeeds or Too-Many-Iterations;
    if Too-Many-Iterations
        then return( No-Solution-Exists )
        else return( Solution );
```

The conclusion that there is no solution is reached with certain probability, hopefully very small, of being wrong. Primality testing using Fermat's little result is a typical example of this type of composition.

References:
[Aho *et al.*, 74], [Bentley *et al.*, 76], [Bentley *et al.*, 78], [Chazelle, 83], [Dwyer, 81], [Lesuisse, 83], [Yao, 77].

2.2.2.2 Alternation

The simplest building operation is alternation. Depending on the result of a test or on the value of a discriminator, one of several alternative algorithms is invoked. For example, based on the value of a command token in a batch updating interpreter, an insertion, modification, or deletion algorithm could be invoked; based on the success of a search in a table, the result could be processed or an error handler called; or based on the size of the input set, an $O(N^2)$ or an $O(N \log N)$ sorting algorithm could be chosen.

There are several forms of alternation that appear in many algorithms; these are elaborated here.

Superimposition
combines two or more algorithms, allowing them to operate on the same data structure more or less independently. Two algorithms F_1 and F_2 may be superimposed over a structure S if $F_1(S,Q_1)$ and $F_2(S,Q_2)$ can both operate together. A typical example of this situation is a file that can be searched by one attribute using F_1 and by another attribute using F_2. Unlike other forms of alternation, the alternative to be used cannot be determined from the state of the structure itself; rather superimposition implies the capability of using any alternative on any instance of the structure involved. Diagrammatically:

Superimposition

---

solve−problem(*A*):
  case 1: $solve{-}problem_1(A)$;
  case 2: $solve{-}problem_2(A)$;
  ...
  case n: $solve{-}problem_n(A)$

Interleaving
is a special case of alternation in which one algorithm does not need to wait for other algorithms to terminate before starting its execution. For example one algorithm might add records to a file while a second algorithm makes deletions; interleaving the two would give an algorithm that performs additions and deletions in a single pass through the file.

Recursion termination
is an alternation that separates the majority of the structure manipulations from the end actions. For example, checking for end of file on input, for reaching a leaf in a search tree, or for reduction to a trivial sublist in a binary search are applications of recursion termination. It is important to realize that this form of alternation is as applicable to iterative processes as recursive ones. Several examples of recursion termination were presented in the previous section on composition (see, for example, divide and conquer).

2.2.2.3 Conformation

If an algorithm builds or changes a data structure, it is sometimes necessary to perform more work to ensure that semantic rules and constraints on the data structure are not violated. For example, when nodes are inserted into or deleted from a tree, the tree's height balance may be altered. As a result it may become necessary to perform some action to restore balance in the new tree. The process of combining an algorithm with a "clean-up" operation on the data structure is called *conformation* (sometimes *organization* or *reorganization).* In effect, conformation is a composition of two algorithms: the original modification algorithm and the constraint satisfaction algorithm. Because this form of composition has an acknowledged meaning to the algorithm's users, it is convenient to list it as a separate class of building operation rather than as a variant of composition. Other examples of conformation include reordering elements in a modified list to restore lexicographic order, percolating newly-inserted elements to their appropriate locations in a priority queue, and removing all dangling (formerly incident) edges from a graph after a vertex is deleted.

2.2.2.4 Self-organization

This is a supplementary heuristic activity that an algorithm may often perform in the course of querying a structure. Not only does the algorithm do its primary work, but it also reaccommodates the data structure in a way designed to improve the performance of future queries. For example, a search algorithm may relocate the desired element once it is found so that future searches through the file will locate the record more quickly. Similarly, a page management system may mark pages as they are accessed, in order that "least recently used" pages may be identified for subsequent replacement.

Once again, this building procedure may be viewed as a special case of composition (or of interleaving); however, its intent is not to build a functionally different algorithm, but rather to augment an algorithm to include improved performance characteristics.

General References:
[Barstow, 80], [Clark *et al.*, 80], [Darlington, 78], [Overmars, 80].

2.2.3 Interchangeability

The framework described so far clearly satisfies two of its goals: it offers sufficient detail to allow effective encoding in any programming language, and it provides a uniformity of description to simplify teaching. It remains to be shown that the approach can be used to discover similarities among implementations as well as to design modifications that result in useful new algorithms.

The primary vehicle for satisfying these goals is the application of interchangeability. Having decomposed algorithms into basic operations used in simple combinations, one is quickly led to the idea of replacing any component of an algorithm by something similar.

The simplest form of interchangeability is captured in the static objects' definition. The hyperrules emphasize similarities among the data structure implementations by indicating the universe of uniform substitutions that can be applied. For example, in any structure using a sequence of reals, the hyperrule for **s−D** together with that for **D** indicates that the sequence of reals can be replaced by a sequence of integers, a sequence of binary trees, etc. Algorithms that deal with such modified structures need at most superficial changes for manipulating the new sequences, although more extensive modifications may be necessary in parts that deal directly with the components of the sequence.

The next level of interchangeability results from the observation that some data structure implementations can be used to simulate the behaviour of others. For example, wherever a bounded sequence is used in an algorithm, it may be replaced by an array, relying on the sequentiality of the integers to access the array's components in order. Sequences of unbounded length may be replaced by sequences of arrays, a technique that may be applied to adapt an algorithm designed for a one-level store to operate in a two-level memory environment wherein each block will hold one array. This notion of interchangeability is the one usually promoted by researchers using abstract data types; their claim is that the algorithms should have been originally specified in terms of abstract sequences. We feel that the approach presented here does not contradict those claims, but rather that many algorithms already exist for specific representations and that an operational approach to specifying algorithms together with the notion of interchangeability is more likely to appeal to data structure practitioners. In cases where data abstraction has been applied, this form of interchangeability can be captured in a metaproduction, as was done for **DICT** in Figure 2.1.

One of the most common examples of this type of interchange is the implementation of linked lists and trees using arrays. For example, an **s−D** is implemented as an $\{\mathbf{D,int}\}_1^N$ and a **bt−D−nil** as an $\{\mathbf{D,int,int}\}_1^N$. In both cases the integers play the same role as the pointers: they select a record of the set. The only difference is syntactic, example

---

$p\uparrow.next \rightarrow next[p]$
$p\uparrow.right \rightarrow right[p]$

Typically the value 0 is reserved to simulate a null pointer.

The most advanced form of interchangeability has not been captured by previous approaches. There are classes of operations that have similar intent yet behave very differently. As a result, replacing some operations by others in the same class may produce startling new algorithms with desirable properties. Some of these equivalence classes are listed below.

Basic Algorithms	{hashing; interpolation; direct addressing}
	{collision resolution methods}
	{binary partition; Fibonaccian partition; median partition; mode partition}
Semantic Rules	{height balance; weight balance}
	{lexicographical order; priority queues}
	{ordered hashing; Brent's hashing; binary tree hashing}
	{minimax; minave}

Chapter 3: SEARCHING ALGORITHMS

3.1 Sequential search

3.1.1 Basic Sequential search

This very basic algorithm is also known as the Linear search or search by Brute Force. It searches for a given element in an array or list by looking through the records sequentially until it finds the element or reaches the end of the structure. Let n denote the size of the array or list on which we search. Let A_n be a random variable representing the number of comparisons made between keys during a successful search and let A_n' be a random variable for the number of comparisons in an unsuccessful search. We have

$$Pr\{A_n = i\} = \frac{1}{n} \qquad (1 \le i \le n)$$

$$E[A_n] = \frac{n+1}{2}$$

$$\sigma^2(A_n) = \frac{n^2-1}{12}$$

$$A_n' = n \, .$$

Below we give pseudo-code descriptions of the Sequential search algorithm in several different situations. The first algorithm searches an array $r[i]$ for the first occurrence of a record with the required key; this is known as primary key search. The second algorithm also searches through an array, but does not stop until it has found every occurrence of the desired key; this is known as secondary key search. The last two algorithms deal with the search for primary and secondary keys in linked lists.

Sequential search in arrays (non-repeated keys)

```
i := 1;
while ( i<n ) and ( key ≠ r[i].k ) do i := i+1;
if key = r[i].k then found( r[i] )
              else   notfound( key );
```

For a faster inner loop, if we are allowed to modify location n+1, then

Sequential search in arrays (non-repeated keys)

```
r[n+1].k := key;
i := 1;
while key ≠ r[i].k do i := i+1;
if i ≤ n then found( r[i] )
        else    notfound( key );
```

Sequential search in arrays (secondary keys)

```
for i:=1 to n do
        if key = r[i].k then found( r[i] );
```

Sequential search in lists (non-repeated keys)

```
for ( p=list; p ≠ NULL && key ≠ p→k; p = p→next );

if ( p == NULL )  notfound( key );
        else   found( *p );
```

Sequential search in lists (secondary keys)

```
p := list;
while p ≠ nil do
        begin
        if key = p↑.k then found( p↑ );
        p := p↑.next
        end;
```

The Sequential search is the simplest search algorithm. Although it is not very efficient in terms of the average number of comparisons needed to find a record, we can justify its use in the following cases:

a) when our files only contain a few records (say, $n \le 20$);

b) when the search will be performed only infrequently;

c) when we are looking for secondary keys and a large number of hits ($O(n)$) is expected;

d) testing extremely complicated conditions.

The Sequential search can also look for a given range of keys instead of one unique key, at no significant extra cost. Another advantage of this search algorithm is that it imposes no restrictions on the order in which records are stored in the list or array.

The efficiency of the Sequential search improves somewhat when we use it to examine external storage. Suppose each physical I/O operation retrieves b records; we say that b is the *blocking factor* of the file, and we refer to each block of b records as a *bucket*. Assume that there are a total of n records in the external file we wish to search and let $k = \lfloor n/b \rfloor$. If we use E_n as a random variable representing the number of external accesses needed to find a given record, we have

$$E[E_n] = k+1-\frac{kb(k+1)}{2n} \approx \frac{k+1}{2}$$

$$\sigma^2(E_n) = \frac{bk(k+1)}{n}\left[\frac{2k+1}{6}-\frac{kb(k+1)}{4n}\right] \approx \frac{k^2}{12}$$

References:

[Berman *et al.*, 74], [Clark, 76], [Flajolet *et al.*, 79], [Flajolet *et al.*, 79], [Flajolet *et al.*, 80], [Gotlieb *et al.*, 78], [Hansen, 78], [Knuth, 73], [Knuth, 74], [Willard, 82], [Wise, 76].

3.1.2 Self-organizing Sequential search: Move to Front method

This algorithm is basically the Sequential search, enhanced with a simple heuristic method for improving the order of the list or array. Whenever a record is found, that record is moved to the front of the table and the other records are slid back to make room for it. (Note that we only need to move the elements which were ahead of the given record in the table; those elements further on in the table need not be touched.) The rationale behind this procedure is that if some records are accessed more often than others, moving those records to the front of the table will decrease the time for future searches. It is, in fact, very common for records in a table to have unequal probabilities of being accessed; thus, the Move to Front technique may often reduce the average access time needed for a successful search.

We will assume that there exists a probability distribution in which $Pr\{accessing\ key\ K_i\} = p_i$. Further we will assume that the keys are numbered in such a way that $p_1 \ge p_2 \ge \cdots \ge p_n > 0$. With this model we have

$$E[A_n] = C_n = \frac{1}{2} + \sum_{i,j} \frac{p_i p_j}{p_i + p_j}$$

$$\sigma^2(A_n) = (2-C_n)(C_n-1) + 4 \sum_{i<j<k} \frac{p_i p_j p_k}{p_i+p_j+p_k} \left(\frac{1}{p_i+p_j} + \frac{1}{p_i+p_k} + \frac{1}{p_j+p_k} \right)$$

$$A_n' = n$$

$$C_n \le \frac{2n}{n+1} C_n^{Opt} = \frac{2n}{n+1} \sum i p_i < 2\mu_1'$$

If we let $T(z) = \sum_{i=1}^{n} z^{p_i}$ then

$$C_n = \int_0^1 z[T'(z)]^2 \, dz$$

Let $C_n(t)$ be the average number of additional accesses required to find a record, given that t accesses have already been made. Starting at $t=0$ with a randomly ordered table we have

$$| C_n(t) - C_n | = O(n^2/t)$$

Below we give a pseudo-code description of the Move to Front algorithm as it can be implemented to search linked lists. The algorithm is less suited to working with arrays.

Self-organizing (Move to Front) Sequential search (lists)

```
if ( list≠0 && key≠list→k )
        /*** Find record ***/
        for ( p=list; (p→next)≠0; p=p→next )
                if ( key==p→next→k )
                        {/*** Move to front of list ***/
                        q = list;
                        list = p→next;
                        p→next = p→next→next;
                        list→next = q;
                        break;
                        };

if ( list≠0 && key==list→k )    found( *list );
        else    notfound( key );
```

There are more sophisticated heuristic methods of improving the order of a list than the Move to Front technique; however, this algorithm can be recommended as particularly appropriate when we have reasons to suspect that the accessing probabilities for individual records will change with time.

Moreover, the Move to Front approach will quickly improve the organization of a list when the accessing probability distribution is very skewed.

If we can guarantee that the search will be successful we can obtain an efficient array implementation by sliding elements back while doing the search.

When searching a linked list, the Move to Front heuristic is preferable to the Transpose heuristic.

Below we give some efficiency measures for this algorithm when the accessing probabilities follow a variety of distributions.

Zipf's law (harmonic): $p_i = (iH_n)^{-1}$

$$C_n = \frac{1}{2}+\frac{(2n+1)H_{2n}-2(n+1)H_n}{H_n} = \frac{2\ln(2)n}{H_n}-\frac{1}{2}+o(1)$$

Lotka's law: $p_i = (i^2H_n^{(2)})^{-1}$

$$C_n = \frac{3}{\pi}\ln n - 0.00206339... + O(\frac{\ln n}{n})$$

Exponential distribution: $p_i = (1-a)a^{i-1}$

$$C_n = -\frac{2\ln 2}{\ln a}-\frac{1}{2}-\frac{\ln a}{24}-\frac{\ln^3 a}{2880}+O(\ln^5 a)$$

Wedge distribution: $p_i = \dfrac{2(n+1-i)}{n(n+1)}$

$$C_n = H_n\left(\frac{4n+2}{3}-\frac{1}{8n(n+1)}\right)-H_{2n}\left(\frac{4n^2+5n-3}{3n}+\frac{3}{4n(n+1)}\right)+\frac{4n+4}{3}-\frac{13}{12(n+1)}$$

$$= \frac{4(1-\ln 2)}{3}n - H_n + \frac{5(1-\ln 2)}{3}+\frac{H_n}{n}+O(n^{-1})$$

Generalized Zipf's: $p_i \propto i^{-\lambda}$

$$C_n \le \frac{1}{\lambda}\left(\psi(\frac{\lambda+1}{2\lambda})-\psi(\frac{1}{2\lambda})\right) \le \frac{\pi}{2}\mu_1'$$

Table 3.1 gives the relative efficiency of Move to Front compared to the optimal arrangement of keys, when the list elements have accessing probabilities which follow several different "folklore" distributions.

References:

[Bitner, 79], [Gonnet *et al.*, 79], [Gonnet *et al.*, 81], [Hendricks, 76], [Knuth, 73], [McCabe, 65], [Rivest, 76], [Tenenbaum *et al.*, 82].

	C_n/Opt			
n	Zipf's law	80%-20% rule	Bradford's law ($b=3$)	Lotka's law
5	1.1921	1.1614	1.1458	1.2125
10	1.2580	1.2259	1.1697	1.2765
50	1.3451	1.3163	1.1894	1.3707
100	1.3623	1.3319	1.1919	1.3963
500	1.3799	1.3451	1.1939	1.4370
1000	1.3827	1.3468	1.1942	1.4493
10000	1.3858	1.3483	1.1944	1.4778

Table 3.1: Relative efficiency of Move to Front

3.1.3 Self-organizing Sequential search: Transpose method

This is another algorithm based on the basic Sequential search and enhanced by a simple heuristic method of improving the order of the list or array. In this model, whenever a search succeeds in finding a record, that record is transposed with the record that immediately precedes it in the table (provided of course that the record being sought was not already in the first position). As with the Move to Front technique, the object of this rearrangement process is to improve the average access time for future searches by moving the most frequently accessed records closer to the beginning of the table. We have

$$E[A_n] = C_n = Prob(I_n)\sum_{\pi}\left(\left(\prod_{i=1}^{n} p_i^{i-\pi(i)}\right)\sum_{j=1}^{n} p_j \pi(j)\right)$$

where π denotes any permutation of the integers $1,2,...,n$. $\pi(j)$ is the location of the number j in the permutation π, and $Prob(I_n)$ is given by

$$Prob(I_n) = \left(\sum_{\pi}\prod_{i=1}^{n} p_i^{i-\pi(i)}\right)^{-1}$$

This expected value of the number of the accesses to find an element can be written in terms of permanents by

$$C_n = \frac{\sum_{k=1}^{n} perm(P_k)}{perm(P)}$$

where P is a matrix with elements $p_{i,j} = p_i^{-j}$ and P_k is a matrix which is the derivative of P with respect to p_k^{-1}. We can put a bound on this expected value by

$$C_n \leq \frac{2n}{n+1} Opt < 2\mu_1'$$

In general the Transpose method gives better results than the Move to Front technique for stable probabilities. In fact, for all record accessing probability distributions, we have

$$C_n^{Transpose} \leq C_n^{MTF}$$

When we look at the case of the unsuccessful search, however, both methods have the identical result

$$A_n' = n$$

Below we give a pseudo-code description of the Transpose algorithm as it can be applied to arrays. The Transpose method can also be implemented efficiently for lists, using an obvious adaptation of the array algorithm.

Self-organizing (Transpose) Sequential search (arrays)

```
i := 1;
while ( i<n ) and ( r[i].k ≠ key ) do i := i+1;

if key = r[i].k then
        begin
        if i>1 then
                begin
                {*** Transpose with predecessor ***}
                tempr := r[i];
                r[i] := r[i-1];
                r[i-1] := tempr;
                i := i-1
                end;
        found( r[i] )
        end
else    notfound( key );
```

It is possible to develop a better self-organizing scheme by allocating extra storage for counters which record how often individual elements are accessed; however, it is conjectured that the Transpose algorithm is the optimal heuristic organization scheme when allocating such extra storage is undesirable.

It should be noted that the Transpose algorithm may take quite some time to rearrange a randomly ordered table into close to optimal order. In fact, it may take $\Omega(n^2)$ accesses to come within a factor of $1+\varepsilon$ of the final steady state.

Because of this slow adaptation ability, the Transpose algorithm is not recommended for applications where accessing probabilities may change with time.

For sequential searching an array, the Transpose heuristic is preferable over the Move to Front heuristic.

References:
[Bitner, 79], [Gonnet *et al.*, 79], [Gonnet *et al.*, 81], [Hendricks, 76], [Rivest, 76], [Tenenbaum, 78].

3.1.4 Optimal Sequential search

When we know the accessing probabilities for a set of records in advance, and we also know that these probabilities will not change with time, we can minimize the average number of accesses in a Sequential search by arranging the records in order of decreasing accessing probability (so that the most often required record is first in the table, and so on). With this preferred ordering of the records, the efficiency measures for the Sequential search are

$$E[A_n] = \mu_1' = \sum_{i=1}^{n} i p_i$$

$$\sigma^2(A_n) = \sum_{i=1}^{n} i^2 p_i - (\mu_1')^2$$

$$A_n' = n$$

Naturally, these improved efficiencies can only be achieved when the accessing probabilities are known in advance and do not change with time. In practice, this is often not the case.

Further, this ordering requires the overhead of sorting all the keys initially according to access probability.

Once the sorting is done, however, the records do not need reorganization during the actual search procedure.

3.1.5 Jump search

Jump search is a technique for searching a sequential ordered file. This technique is applicable whenever it is possible to jump or skip a number of records at a cost that is substantially less than the sequential read of those records.

Let a be the cost of a jump and b the cost of a sequential search. If the jumps have to be of a fixed size, then the optimum jump size is $\sqrt{na/b}$, and consequently

$$C_n = \sqrt{nab} + O(1)$$

Doing uniform size jumps is not the best strategy. It is better to have larger jumps at the beginning and smaller jumps at the end, so that the average and worst-case searching times are minimized.

For the optimal strategy the ith jump should be of $\sqrt{2an/b} - ai/b$ records; then:

$$C_n = \sqrt{\frac{8abn}{9}} + O(1)$$

Jump search algorithm

```
read-first-record;
while key > r.k do Jump-records;
while key < r.k do read-previous-record;
if key=r.k then     found( r )
            else    notfound( key );
```

This method can be extended to an arbitrary number of levels; at each level the size of the jump is different. Ultimately when we use $\log_2 n$ levels this algorithm coincides with Binary search.

There are two situations in which this algorithm becomes an appealing alternative:

(1) Tape reading where we can command to skip physical records almost without cost to the computer;

(2) Sequential files with compressed and/or encoded information when the cost of decompressing and/or decoding is very high.

When Binary search is possible there is no reason to use Jump searching.

References:

[Janko, 81], [Leipala, 81], [Shneiderman, 78], [Six, 73].

General References:

[Allen *et al.*, 76], [McKellar *et al.*, 78], [Standish, 80], [Wirth, 76].

3.2 Sorted array search

The following algorithms are designed to search for a record in an array whose keys are arranged in order. Without loss of generality we will assume an increasing order.

We will discuss only the searching algorithms. The insertion of new elements or direct construction of a sorted array is the same for all algorithms. These searching algorithms are not efficient when the table undergoes a lot of insertions and deletions. Both updating operations cost $O(n)$ work each. It is then implicit that these tables are rather static.

Insertion into an ordered array

```
i := n;
if n≥m then Error {*** Table full ***}
else begin
        n := n+1;
        while i>0 do
                if r[i].k > key then begin
                        r[i+1] := r[i];
```

```
                i := i-1
                end
         else   goto 999; {*** break ***}

    {*** Insert new record ***}
    999:
    r[i+1].k := key
    end;
```

The above algorithm will not detect the insertion of duplicates, i.e. elements already present in the table. If we have all the elements available at the same time, it is advantageous to sort them in order, as opposed to inserting them one by one.

General References:
[Flajolet *et al.*, 79], [Flajolet *et al.*, 80], [Guibas, 77], [Horowitz *et al.*, 76], [Overholt, 73], [Peterson, 57], [Price, 71].

3.2.1 Binary search

This algorithm is also known as the Bipartition search, the Bisection search, or the Dichotomic search. It searches a sorted array using the "tail recursion" technique. At each step of the search, a comparison is made with the middle element of the array. Then the algorithm decides which half of the array should contain the required key, and discards the other half. The process is repeated, halving the number of records to be searched at each step until only one key is left. At this point one comparison is needed to decide whether the searched key is present in the file. If the array contains n elements and $k = \lfloor \log_2 n \rfloor$ then we have:

$$\lfloor \log_2 (n+1) \rfloor \le A_n = A_n' \le k+1$$

$$C_n = C_n' = k+2-\frac{2^{k+1}}{n+1} \approx \log_2 n$$

$$\sigma^2(A_n') \le \frac{1}{12}$$

If we use three-way comparisons and stop the search on equality, the number of comparisons for the successful search changes to:

$$1 \le A_n \le k+1$$

$$E[A_n] = C_n = k+1-\frac{2^{k+1}-k-2}{n} \approx \log_2(n)-1+\frac{k+2}{n}$$

$$\sigma^2(A_n) = \frac{3\times 2^{k+1}-(k+2)^2-2}{n} - \left(\frac{2^{k+1}-k-2}{n}\right)^2$$

$$\approx 2.125 \pm .125 + o(1)$$

$$C_n = (1+\frac{1}{n})C_n' - 1$$

(The random variables A_n and A_n' are as defined in section 3.1; C_n and C_n' are the expected values of A_n and A_n' respectively.)

Binary search algorithm

```
low := 0;
high := n;
while high-low > 1 do begin
            j := (high+low) div 2;
            if key <= r[j].k then high := j
                         else     low := j
            end;
if key = r[high].k then   found( r[high] )
                 else     notfound( key );
```

There are more efficient search algorithms than the Binary search but such methods must perform a number of special calculations: for example, the Interpolation search (Section 3.2.2) calculates a special interpolation function, while hashing algorithms (Section 3.3) must compute one or more hashing functions. The Binary search is an optimal search algorithm when we restrict our operations only to comparisons between keys.

Binary search is a very stable algorithm: the range of search times stays very close to the average search time, and the variance of the search times is $O(1)$.

Another advantage of the Binary search is that it is well suited to searching for keys in a given range as well as searching for one unique key.

One drawback of the Binary search is that it requires a sorted array. Thus additions, deletions, and modifications to the records in the table can be expensive, requiring work on the scale of $O(n)$.

Table 3.2 gives figures showing the performance of the three-way comparison Binary search for various array sizes.

References:

[Aho *et al.*, 74], [Bentley *et al.*, 76], [Berman *et al.*, 74], [Flajolet *et al.*, 79], [Flores *et al.*, 71], [Gotlieb *et al.*, 78], [Jones, 72], [Knuth, 73], [Leipala, 79], [Lesuisse, 83], [Overholt, 73], [Yao, 81].

n	C_n	$\sigma^2(A_n)$	C_n'
5	2.2000	0.5600	2.6667
10	2.9000	0.8900	3.5455
50	4.8600	1.5204	5.7451
100	5.8000	1.7400	6.7327
500	7.9960	1.8600	8.9780
1000	8.9870	1.9228	9.9770
5000	11.3644	2.2004	12.3619
10000	12.3631	2.2131	13.3618

Table 3.2: Exact results for Binary search

3.2.2 Interpolation search

This is also known as the Estimated Entry search. It is one of the most natural ways to search an ordered table which contains numerical keys. Like the Binary search, it uses the "tail recursion" approach, but in a more sophisticated way. At each step of the search, the algorithm makes a guess (or interpolation) of where the desired record is apt to be in the array, basing its guess on the value of the key being sought and the values of the first and last keys in the table. As with the Binary search, we compare the desired key with the key in the calculated probe position; if there is no match, we discard the part of the file we know does not contain the desired key and probe the rest of the file using the same procedure recursively.

Let us suppose we have normalized the keys in our table to be real numbers in the closed interval [0,1] and let $\alpha \in [0,1]$ be the key we are looking for. For any integer $k \leq n$, the probability of needing more than k probes to find α is given by

$$Pr\{A_n > k\} \approx \prod_{i=1}^{k} (1 - \frac{1}{2}\xi^{2^{-i}})$$

where $\xi = \dfrac{2}{\pi n \alpha(1-\alpha)}$.

$$E[A_n] = \log_2 \log_2 n + O(1)$$
$$\approx \log_2 \log_2 (n+3)$$
$$\sigma^2(A_n) = O(\log_2 \log_2 n)$$
$$E[A_n'] = \log_2 \log_2 n + O(1)$$
$$\approx \log_2 \log_2 n + 0.58$$

When implementing the Interpolation search, we must make use of an interpolating formula. This is a function $\phi(\alpha, n)$ which takes as input the desired key $\alpha (\alpha \in [0,1])$ and the array of length n, and which yields an array index between 1 and n, essentially a guess at where the desired array element is. Two of the simplest linear interpolation formulas are $\phi(\alpha, n) = \lceil n\alpha \rceil$ and $\phi(\alpha, n) = \lfloor n\alpha + 1 \rfloor$. Below we give a description of the Interpolation search.

Interpolation search algorithm

```
low := 1;
high := n;
while (r[high].k ≥ key) and (key > r[low].k) do
        begin
        j := trunc( (key−r[low].k) / (r[high].k−r[low].k) *
                                (high−low) ) + low;
        if     key > r[j].k then low  := j+1
        else if key < r[j].k then high := j−1
                else              low  := j
        end;
if key = r[low].k then      found( r[low] )
                else        notfound( key );
```

The Interpolation search is asymptotically optimal among all algorithms which search arrays of numerical keys. However, it is very sensitive to a non-uniform [0,1] distribution of the keys. Simulations show that the Interpolation search can lose its $O(\log \log n)$ behaviour under some non-uniform key distributions.

While it is relatively simple in theory to adjust the algorithm to work suitably even when keys are not distributed uniformly, difficulties can arise in practice. First of all, it is necessary to know how the keys are distributed and this information may not be available. Furthermore, unless the keys follow a very simple probability distribution, the calculations required to adjust the algorithm for non-uniformities can become quite complex and hence impractical.

Interpolation search will not work if key values are repeated.

Table 3.3 gives figures for the efficiency measures of the Interpolation search for various array sizes. The most important cost in the algorithm is the computation of the interpolation formula. For this reason, we will count the number of times the body of the while-loop is executed (A_n). The amount L_n is the average of the worst-case A_n for every file.

References:
[Burkhard, 83], [Burton *et al.*, 80], [Ehrlich, 81], [Franklin, 79], [Gonnet *et al.*, 80], [Gonnet, 77], [Gotlieb *et al.*, 78], [Kruijer, 74], [Lewis *et al.*, 81], [Nat, 79], [Perl *et al.*, 77], [Perl *et al.*, 78], [Reif, 82], [Waters, 75], [Whitt *et al.*, 75], [Yao *et al.*, 76].

3.2.3 Interpolation-Sequential search

This algorithm is a combination of the Interpolation and Sequential search methods. An initial interpolation probe is made into the table, just as in the Interpolation algorithm; if the given element is not found in the probed

n	$E[A_n]$	L_n	$E[A_n']$
5	0.91551 ± 0.00071	1.4524 ± 0.0015	1.28003 ± 0.00065
10	1.25164 ± 0.00088	2.1854 ± 0.0025	1.50407 ± 0.00087
50	1.9162 ± 0.0011	3.8252 ± 0.0081	2.0273 ± 0.0012
100	2.1519 ± 0.0012	4.556 ± 0.012	2.2403 ± 0.0012
500	2.6083 ± 0.0013	6.180 ± 0.027	2.6710 ± 0.0013
1000	2.7759 ± 0.0013	6.838 ± 0.036	2.8382 ± 0.0013
5000	3.1052 ± 0.0013	8.215 ± 0.077	3.1636 ± 0.0013
10000	3.2214 ± 0.0013	8.74 ± 0.10	3.2786 ± 0.0013

Table 3.3: Simulation results for Interpolation search

position, the algorithm then proceeds to search through the table sequentially, forwards or backwards depending on which direction is appropriate. Let A_n and A_n' be random variables representing the number of array accesses for successful and unsuccessful searches respectively. We have

$$E[A_n] = 1 + \frac{2}{n}\sum_{k=1}^{n-1} \frac{\Gamma(n)}{\Gamma(k)\Gamma(n-k)} (k/n)^k (1-k/n)^{n-k}$$

$$= 1 + \left(\frac{n\pi}{32}\right)^{1/2} (1 - \frac{7}{12n}) + O(n^{-1})$$

$$E[A_n'] = \frac{2}{n+1}\left(1 + \sum_{k=1}^{n-1} \frac{(n+1)k}{n} I_{k/n}(k+1, n-k) - (k+1) I_{k/n}(k+2, n-k)\right)$$

$$= \left(\frac{n\pi}{32}\right)^{1/2} + O(1)$$

where $I_x(a,b)$ denotes the incomplete beta function.

As with the standard Interpolation search (Section 3.2.2), this method requires an interpolation formula ϕ such as $\phi(\alpha,n) = \lceil n\alpha \rceil$ or $\phi(\alpha,n) = \lfloor n\alpha + 1 \rfloor$; for the code below we use the latter.

Interpolation-sequential search

```
if n > 1 then
      begin
      {*** initial probe location ***}
      j := trunc( (key-r[1].k) / (r[n].k-r[1].k) ) + 1;
      if key < r[j].k then
                  while (j>1) and (key<r[j].k) do j := j-1
            else  while (j<n) and (key>r[j].k) do j := j+1
      end
else  j := 1;

if key = r[j].k then found( r[j] )
```

```
else        notfound( key );
```

Asymptotically, this algorithm behaves significantly worse than the pure Interpolation search. Note however, that for $n<500$ it is still more efficient than Binary search.

When we use this search technique with external storage, we have a significant improvement over the internal case. Suppose we have storage buckets of size b (that is, each physical I/O operation reads in a block of b records); then the number of external accesses the algorithm must make to find a record is given by

$$E[E_n] = 1 + \frac{1}{b}\left(\frac{n\pi}{32}\right)^{1/2} + O(n^{-1/2})$$

In addition to this reduction the accessed buckets are contiguous and hence the seek time may be reduced.

Table 3.4, below, lists the expected number of accesses required for both successful and unsuccessful searches for various table sizes.

n	$E[A_n]$	$E[A_n']$
5	1.5939	1.9613
10	1.9207	2.3776
50	3.1873	3.7084
100	4.1138	
500	7.9978	
1000	10.9024	
5000	23.1531	
10000	32.3310	

Table 3.4: Exact results for Interpolation Sequential search

References:
[Gonnet *et al.*, 77].

3.3 Hashing

Hashing or scatter storage algorithms are distinguished by the use of a *hashing function.* This is a function which takes a key as input and yields an integer in a prescribed range (e.g. $[0,m-1]$) as a result. The function is designed so that the integer values it produces are uniformly distributed throughout the range. These integer values are then used as indices for an array of size m called the *hashing table.* Records are both inserted into the table and retrieved from the table by using the hashing function to calculate the required indices from the record keys.

When the hashing function yields the same index value for two different keys, we have a *collision.* A complete hashing algorithm consists of a hashing function and a method for handling the problem of collisions. Such a method is called a *collision resolution scheme.*

There are two distinct classes of collision resolution schemes. The first class is called *open addressing.* Schemes in this class resolve collisions by computing new indices based on the value of the key; in other words, they "rehash" into the table. In the second class of resolution schemes, all elements which "hash" to the same table location are linked together in a chain.

To insert a key using open-addressing we follow a sequence of probes in the table. This sequence of probe positions is called a *path.* In open addressing a key will be inserted in the first empty location of its path. There are at most $m!$ different paths through a hashing table and most open-addressing methods use far less paths than $m!$ Several keys may share a common path or portions of a path. The portion of a path which is fully occupied with keys will be called a *chain.*

The undesirable effect of having chains longer than expected is called *clustering.* More precisely there are two possible definitions for clustering.

a) Let $p = \Theta(m^k)$ be the maximum number of different paths. We say that a collision resolution scheme has $k+1$ clustering if it allows p different circular paths. A *circular path* is the set of all paths that are obtained from circular permutations of a given path. In other words, all the paths in a circular path share the same order of table probing except for their starting position.

b) If the path depends exclusively on the first k initial probes we say that we have k-clustering.

It is generally agreed that linear probing suffers from primary clustering, quadratic and double hashing from secondary clustering, and uniform and random probing from no clustering.

Assume our hashing table of size m has n records stored in it. The quantity $\alpha = n/m$ is called the *load factor* of the table. We will let A_n be a random variable which represents the number of times a given algorithm must access the hashing table to locate any of the n elements stored there. It is expected that some records will be found on the first try, while for others we may have to rehash several times or follow a chain of other records before we locate the record we want. We will use L_n to denote the length of the longest probe sequence needed to find any of the n records stored in the table. Thus our random variable A_n will have the range

$$1 \leq A_n \leq L_n$$

Its actual value will depend on which of the n records we are looking for.

In the same way, we will let A_n' be a random variable which represents the number of accesses required to insert an $n+1^{st}$ element into a table already containing n records. We have

$$1 \leq A_n' \leq n+1$$

The search for a record in the hashing table starts at an initial probe location calculated by the hashing function and from there follows some prescribed sequence of accesses determined by the algorithm. If we find an empty location in the table while following this path, we may conclude that the desired record is not in the file. Thus it is important that an open addressing scheme be able to tell the difference between an empty table position (one that has not yet been allocated) and a table position which has had its record deleted. The probe sequence may very well continue past a deleted position, but an empty position marks the end of any search. When we are inserting a record into the hashing table rather than searching for one, we use the first empty or deleted location we find.

Let

$$C_n = E[A_n]$$

and

$$C_n' = E[A_n']$$

C_n denotes the expected number of accesses needed to locate any individual record in the hashing table while C_n' denotes the expected number of accesses needed to insert a record. Thus

$$C_n = \frac{1}{n}\sum_{i=0}^{n-1} E[A_i'] = \frac{1}{n}\sum_{i=0}^{n-1} C_i'$$

Due to their simplicity we will describe the hash table algorithms as in-line code rather than as functions. In all cases we will search in an array of records named r, with definition:

Search array definition

```
r : array [ 0..m-1 ] of record
        k : typekey;
        . . . .
        end;
```

The key value being searched is stored in the variable *key*. There exist functions (or default values) that indicate whether an entry is empty (*empty*($r[i]$)) or indicate whether a value has been deleted (*deleted*($r[i]$)). The hashing functions yield values between 0 and $m-1$. The increment functions, used for several double-hashing algorithms, yield values between 1 and $m-1$.

General References:

[Aho *et al.*, 74], [Aho *et al.*, 83], [Anderson *et al.*, 79], [Bays, 73], [Bobrow, 75], [Bolour, 79], [Bolour, 81], [Burkhard, 79], [Carter *et al.*, 79], [Cichelli, 80], [Clapson, 77], [Deutscher *et al.*, 75], [Devillers *et al.*, 79], [Dubost *et al.*, 75], [Fredman *et al.*, 82], [Gotlieb *et al.*, 78], [Guibas, 76], [Guibas, 78], [Halatsis *et al.*, 78], [Horowitz *et al.*, 76], [Jaeschke, 81], [Knott, 75], [Knuth, 73], [Larson, 80], [Lipton *et al.*, 80], [Litwin, 81], [Lum *et al.*, 71], [Lum *et al.*, 72], [Lum, 73], [Mairson, 83], [Markowsky *et al.*, 78], [Maurer *et al.*, 75], [Maurer, 68], [Mehlhorn, 82], [Mendelson *et al.*, 79], [Papadimitriou *et al.*, 80], [Peterson, 57], [Pippenger, 79], [Rivest, 74], [Rosenberg *et al.*, 77], [Rosenberg, 80], [Rothnie *et al.*, 74], [Sarwate, 80], [Scheuermann, 79], [Severance *et al.*, 76], [Sprugnoli, 77], [Standish, 80], [Tai *et al.*, 80], [Trabb Pardo, 78], [Ullman, 72], [Wegman *et al.*, 79].

3.3.1 Uniform probing hashing

Uniform probing hashing is an open addressing scheme which resolves collisions by probing the table according to a permutation of the integers $[1,m]$. The permutation used depends only on the key of the record in question. Thus for each key, the order in which the table is probed is a random permutation of all table locations. This method will equally likely use any of the $m!$ possible paths.

Uniform probing is a theoretical hashing model which has the advantage of being relatively simple to analyze. The following list summarizes some of the pertinent facts about this scheme:

$$Pr\{A_n' > k\} = \frac{n^{\underline{k}}}{m^{\underline{k}}}$$

where $n^{\underline{k}}$ denotes the descending factorial, i.e. $n^{\underline{k}} = n(n-1)\ldots(n-k+1)$.

$$E[A_n] = C_n = \frac{m+1}{n}[H_{m+1} - H_{m-n+1}] \approx -\alpha^{-1}\ln(1-\alpha)$$

$$\sigma^2(A_n) = \frac{2(m+1)}{m-n+2} - C_n(C_n+1)$$

$$\approx \frac{2}{1-\alpha} + \alpha^{-1}\ln(1-\alpha) - \alpha^{-2}\ln^2(1-\alpha)$$

$$C_m = \frac{m+1}{m}[H_{m+1} - 1] = \ln m + \gamma - 1 + o(1)$$

$$C_n^{worst\ file} = \frac{n+1}{2}$$

$$E[A_n'] = C_n' = \frac{m+1}{m-n+1} \approx \frac{1}{1-\alpha}$$

$$\sigma^2(A_n') = \frac{(m+1)n(m-n)}{(m-n+1)^2(m-n+2)} \approx \frac{\alpha}{(1-\alpha)^2}$$

$$C_m' = m$$

$$C_n'^{\,worst\ file} = C_n' = \frac{m+1}{m-n+1}$$

$$L_n = \max_{0 \le i < n} A_i'$$

$$1 \le L_n \le n$$

$$E[L_n] = -\log_\alpha m - \log_\alpha(-\log_\alpha m) + O(1)$$

$$E[L_m] = 0.631587... \times m + O(1)$$

$$E[keys\ requiring\ i\ accesses] = \frac{n^{\underline{i}}}{m^{\underline{i}}}\left(\frac{(m-n)i+m+1}{i(i+1)}\right)$$

Table 3.5 gives figures for some of the quantities we have been discussing in the cases $m=100$ and $m=\infty$.

	$m=100$			$m=\infty$		
α	C_n	$\sigma^2(A_n)$	C_n'	C_n	$\sigma^2(A_n)$	C_n'
50%	1.3705	0.6358	1.9804	1.3863	0.6919	2.0
80%	1.9593	3.3837	4.8095	2.0118	3.9409	5.0
90%	2.4435	8.4190	9.1818	2.5584	10.8960	10.0
95%	2.9208	17.4053	16.8333	3.1534	26.9027	20.0
99%	3.7720	44.7151	50.0	4.6517	173.7101	100.0

Table 3.5: Exact results for Uniform probing hashing

It does not seem practical to implement a clustering-free hashing function.

Double hashing behaves very similarly to Uniform probing. For all practical purposes they are indistinguishable.

References:
[Ajtai *et al.*, 78], [Gonnet, 80], [Gonnet, 81], [Knuth, 73], [Larson, 80], [Larson, 83].

3.3.2 Random probing hashing

This is an open addressing hashing scheme in which collisions are resolved by additional probes into the table. The sequence of these probes is considered to be random and depends only on the value of the key. The difference between this scheme and Uniform probing is that here some positions may be repeated in the probe sequence, whereas in Uniform probing no position is examined more than once. Random probing is another theoretical model which is relatively simple to analyze.

The pertinent formulas for this scheme are given by:

$$Pr\{A_n' > k\} = \alpha^k$$

$$E[A_n] = C_n = \frac{m}{n}[H_m - H_{m-n}] = -\alpha^{-1}\ln(1-\alpha) + O(\frac{1}{m-n})$$

$$\sigma^2(A_n) = \frac{2m^2}{n}[H_m^{(2)} - H_{m-n}^{(2)}] - C_n(C_n+1)$$

$$= \frac{2}{1-\alpha} + \alpha^{-1}\ln(1-\alpha) - \alpha^{-2}\ln^2(1-\alpha) + O(\frac{1}{m-n})$$

$$C_n^{worst\ file} = \infty$$

$$1 \leq A_n' \leq \infty$$

$$E[A_n'] = C_n' = \frac{1}{1-\alpha}$$

$$\sigma^2(A_n') = \frac{\alpha}{(1-\alpha)^2}$$

$$C_m = H_m = \ln m + \gamma + O(m^{-1})$$

Table 3.6 gives figures for some of the basic complexity measures in the case of $m=100$ and $m=\infty$.

	$m=100$			$m=\infty$		
α	C_n	$\sigma^2(A_n)$	C_n'	C_n	$\sigma^2(A_n)$	C_n'
50%	1.3763	0.6698	2.0	1.3863	0.6919	2.0
80%	1.9870	3.7698	5.0	2.0118	3.9409	5.0
90%	2.5093	10.1308	10.0	2.5584	10.8960	10.0
95%	3.0569	23.6770	20.0	3.1534	26.9027	20.0
99%	4.2297	106.1598	100.0	4.6517	173.7101	100.0

Table 3.6: Exact results for Random probing hashing

Notice that the asymptotic results ($m \rightarrow \infty$; α fixed) coincide with Uniform probing, while for finite values of m, Uniform probing gives better results.

Random probing could be implemented using pseudo-random probe locations; it does not seem, however, to be a good alternative to the Double hashing algorithm described in section 3.3.4.

References:
[Larson, 82], [Morris, 68].

3.3.3 Linear probing hashing

Linear probing is an open addressing hashing algorithm that resolves collisions by probing to the next table location modulo m. In other words, it probes sequentially through the table starting at the initial hash index, possibly running until it reaches the end of the table, rolling to the beginning of the table if necessary, and continuing the probe sequence from there. This method resolves collisions using only one circular path. For this model:

$$E[A_n] = C_n = \frac{1}{2}\left(1 + \sum_{k \geq 0} \frac{(n-1)^{\underline{k}}}{m^k}\right)$$

$$= \frac{1}{2}\left(1 + \frac{1}{1-\alpha}\right) - \frac{1}{2(1-\alpha)^3 m} + O(m^{-2})$$

$$\sigma^2(A_n) = \frac{1}{6} + \frac{m}{n}\left(\sum_{k \geq 0} \frac{k^2+k+3}{6} \frac{n^{\underline{k}}}{m^k} - \frac{1}{2}\right) - (C_n)^2$$

$$= \frac{\alpha(\alpha^2-3\alpha+6)}{12(1-\alpha)^3} - \frac{3\alpha+1}{2(1-\alpha)^5 m} + O(m^{-2})$$

$$C_n^{(worst\ file)} = \frac{n+1}{2}$$

$$C_m = \sqrt{\pi m/8} + \frac{1}{3} + O(m^{-1/2})$$

$$E[A_n'] = C_n' = \frac{1}{2}\left(1+\sum_{k\ge 0}\frac{(k+1)n^{\underline{k}}}{m^k}\right)$$

$$= \frac{1}{2}\left(1+\frac{1}{(1-\alpha)^2}\right) - \frac{3\alpha}{2(1-\alpha)^4 m} + O(m^{-2})$$

$$\sigma^2(A_n') = \frac{1}{6} + \sum_{k\ge 0}\frac{(k+1)(k^2+3k+5)}{12}\frac{n^{\underline{k}}}{m^k} - (C_n')^2$$

$$= \frac{3(1-\alpha)^{-4}}{4} - \frac{2(1-\alpha)^{-3}}{3} - \frac{1}{12} - \frac{\alpha(8\alpha+9)}{2(1-\alpha)^6 m} + O(m^{-2})$$

$$C_n'^{(worst\ file)} = 1+\frac{n(n+1)}{2m}$$

We denote the hashing table as an array r, with each element $r[i]$ having a key k.

Linear probing hashing: search

```
i := hashfunction( key ) ;
last := (i+n-1) mod m;
while (i≠last) and (not empty(r[i])) and (r[i].k≠key) do
        i := (i+1) mod m;
if r[i].k=keythen found(r[i])
                else notfound(key);
```

Linear probing hashing: insertion

```
i := hashfunction( key ) ;
last := (i+m-1) mod m;
while (i≠last) and (not empty(r[i]))
        and (not deleted(r[i])) and (r[i].k≠key) do
                i := (i+1) mod m;
if empty(r[i]) or deleted(r[i]) then
                begin
                {*** insert here ***}
```

```
            r[i].k := key;
            n := n+1
            end
    else Error {*** table full, or key already in table ***};
```

Linear probing hashing uses one of the simplest collision resolution techniques available, requiring a single evaluation of the hashing function. It suffers, however, from a piling-up phenomenon called primary clustering. The longer a contiguous sequence of keys grows, the more likely it is that collisions with this sequence will occur when new keys are added to the table. Thus the longer sequences grow faster than the shorter ones. Furthermore, there is a greater probability that longer chains will coalesce with other chains, causing even more clustering. This problem makes the Linear probing scheme undesirable with a high load factor α.

It should be noted that the number of accesses in a successful or unsuccessful search has a very large variance. Thus it is possible that there will be a sizable difference in the number of accesses needed to find different elements.

It should also be noted that given any set of keys, the order in which the keys are inserted has no effect on the total number of accesses needed to install the set.

An obvious variation on the Linear probing scheme is to move backward through the table instead of forward, when resolving collisions. More generally, we could move through a unique permutation of the table entries, which would be the same for every key; only the starting point of the permutation would depend on the key in question. Clearly, both these variations would exhibit exactly the same behaviour as the standard Linear probing model.

As noted previously, deletions from the table must be marked as such in order for the algorithm to work correctly. The presence of deleted records in the table is called contamination, a condition which clearly interferes with the efficiency of an unsuccessful search. When new keys are inserted after deletions, the successful search is also deteriorated.

Up until now, we have been considering the shortcomings of Linear probing when it is used to access internal storage. With external storage, the performance of the scheme improves significantly, even for fairly small storage buckets. Let b be the blocking factor, i.e. the number of records per storage bucket. We find that the number of external accesses (E_n) is

$$E_n = 1 + \frac{A_n - 1}{b}$$

while the number of accesses required to insert an $n+1^{st}$ record is

$$E_n' = 1 + \frac{A_n' - 1}{b}$$

Furthermore, for external storage, we may change the form of the algorithm so that we scan each bucket completely before examining the next bucket. This improves the efficiency somewhat over the simplest form of the Linear probing algorithm.

Table 3.7 gives figures for the efficiency of the Linear probing scheme with $m = 100$, and $m = \infty$.

	$m=100$			$m=\infty$		
α	C_n	$\sigma^2(A_n)$	C_n'	C_n	$\sigma^2(A_n)$	C_n'
50%	1.4635	1.2638	2.3952	1.5	1.5833	2.5
80%	2.5984	14.5877	9.1046	3.0	35.3333	13.0
90%	3.7471	45.0215	19.6987	5.5	308.25	50.5
95%	4.8140	87.1993	32.1068	10.5	2566.58	200.5
99%	6.1616	156.583	50.5	50.5	330833.	5000.5

Table 3.7: Exact results for Linear probing hashing

References:
[Bandyopadhyay, 77], [Blake *et al.*, 77], [Knuth, 73], [Konheim *et al.*, 66], [Larson, 79], [Larson, 82], [Lyon, 78], [Mendelson *et al.*, 80], [Morris, 68], [Quittner *et al.*, 81], [Samson, 81], [Schay *et al.*, 62].

3.3.4 Double hashing

Double hashing is an open addressing hashing algorithm which resolves collisions by means of a second hashing function. This second function is used to calculate an increment less than m which is added on to the index to make successive probes into the table. Each different increment gives a different path, hence this method uses $m-1$ circular paths. We have

$$E[A_n] = C_n = -\alpha^{-1}\ln(1-\alpha)+o(1) \quad (\alpha<0.319...)$$

$$E[A_n'] = C_n' = (1-\alpha)^{-1}+o(1) \quad (\alpha<0.319...)$$

Actually, Double hashing is not identical to Uniform probing. E.g. if $m=13$ then

$$C_{13}^{Doub.\ hash.} - C_{13}^{Unif.\ prob.} = 0.0009763...$$

$$E[L_{13}^{Doub.\ hash.}] - E[L_{13}^{Unif.\ prob.}] = 0.001371...$$

Below we give descriptions of search and insertion algorithms which implement the Double hashing scheme. Both algorithms require the table size m to be a prime number; otherwise there is the possibility that the probe sequence, for some keys, will not cover the entire table.

Double hashing: search

```
i := hashfunction( key ) ;
inc := increment( key ) ;
last := (i+(n-1)*inc) mod m;
while (i≠last) and (not empty(r[i])) and (r[i].k≠key) do
        i := (i+inc) mod m;
if r[i].k=keythen found(r[i])
              else notfound(key);
```

Double hashing: insertion

```
i := hashfunction( key ) ;
inc := increment( key );
last := (i+(m-1)*inc) mod m;
while (i≠last) and (not empty(r[i]))
      and (not deleted(r[i])) and (r[i].k≠key) do
              i := (i+inc) mod m;
if empty(r[i]) or deleted(r[i]) then
              begin
              {*** insert here ***}
              r[i].k := key;
              n := n+1
              end
else Error {*** table full, or key already in table ***};
```

Double hashing is a practical and efficient hashing algorithm. Since the increment we use to step through the table depends on the key we are searching for, Double hashing does not suffer from primary clustering. This also implies that changing the order of insertion of a set of keys may change the average number of accesses required to do the inserting. Thus several reorganization schemes have been developed to reorder insertion of keys in ways which make Double hashing more efficient.

As with Linear probing, deletion of records leads to contamination and decreases the efficiency of the unsuccessful search. When new keys are inserted after deletions, the successful search is also deteriorated. The unsuccessful search can be improved by keeping in a counter the length of the longest probe sequence (llps) in the file. Thus the search algorithm is the same as before, except that the variable *last* is computed as

```
last := (i+(llps-1)*inc) mod m;
```

Whenever we insert a new key we may need to update this counter.

Extensive simulations show that it is practically impossible to establish statistically whether Double hashing behaves differently from Uniform probing. For example we would need a sample of 3.4×10^7 files of size 13 to show statistically with 95% confidence that Double hashing is different from Uniform probing. The following tables list some sample results.

	$m=101$			
n	C_n	$\sigma^2(A_n)$	L_n	C'_n
51	1.3772±0.0011	0.6553±0.0036	4.585±0.014	2.0015±0.0013
81	1.9707±0.0026	3.485±0.022	11.054±0.045	4.8629±0.0051
91	2.4572±0.0043	8.666±0.065	18.185±0.089	9.316±0.011
96	2.9361±0.0060	17.80±0.14	27.12±0.14	17.020±0.019
100	3.7877±0.0096	50.22±0.43	48.90±0.23	51.0

Table 3.8: Simulation results for Double hashing

	$m=4999$			
n	C_n	$\sigma^2(A_n)$	L_n	C'_n
2500	1.3855±0.0012	0.6885±0.0039	9.30±0.11	2.0012±0.0014
3999	2.0129±0.0028	3.943±0.028	25.72±0.42	5.0030±0.0056
4499	2.5574±0.0046	10.867±0.093	48.39±0.93	9.974±0.012
4749	3.1464±0.0072	26.63±0.30	88.5±2.0	19.969±0.026
4949	4.614±0.018	163.3±3.3	313.7±8.7	98.14±0.13

Table 3.9: Simulation results for Double hashing

References:
[Bell *et al.*, 70], [Guibas *et al.*, 78], [Guibas, 76], [Knuth, 73], [Luccio, 72], [Samson, 81].

3.3.5 Quadratic hashing

Quadratic hashing is an open addressing algorithm that resolves collisions by probing the table in increasing increments modulo the table size, i.e. $h(k)$, $h(k)+1$, $h(k)+3$, $h(k)+6$,.... If the increments are considered to be a random permutation of the integers $1...m$, we obtain the results

$$E[A_n] = C_n = \frac{n+1}{n}(H_{m+1}-H_{m-n+1})+1-\frac{n}{2(m+1)}+O(n^{-1})$$

$$\approx 1-\ln(1-\alpha)-\frac{\alpha}{2}$$

$$E[A'_n] = C'_n = \frac{m+1}{m-n+1}-\frac{n}{m+1}+H_{m+1}-H_{m-n+1}+O(m^{-1})$$

$$\approx (1-\alpha)^{-1}-\alpha-\ln(1-\alpha)$$

Quadratic hashing: search

```
i := hashfunction( key ) ;
inc := 0;
while (inc<m) and (not empty(r[i])) and (r[i].k≠key) do
        begin
        i := (i+inc+1) mod m;
        inc := inc + 2
        end;
if r[i].k=key then found(r[i])
                else notfound(key);
```

Quadratic hashing: insertion

```
i := hashfunction( key ) ;
inc := 0;
while (inc<m) and (not empty(r[i])) and
        (not deleted(r[i])) and (r[i].k≠key) do begin
                i := (i+inc+1) mod m;
                inc := inc + 2;
                end;
if empty(r[i]) or deleted(r[i]) then
                begin
                {*** insert here ***}
                r[i].k := key;
                n := n+1
                end
else Error {*** table full, or key already in table ***};
```

Quadratic hashing requires a single hashing function evaluation per search. It suffers, however, from a slight piling-up phenomenon called secondary clustering.

This algorithm may fail to insert a key after the table is half full. This is due to the fact that the ith probe coincides with the $m-i^{th}$ probe. This can be solved by the use of the probe sequence $h(k)$, $h(k)+1$, $h(k)-1$, $h(k)+4$, $h(k)-4$, ...

The following tables show some simulation results for Quadratic hashing. F_n indicates the average number of times that the algorithm failed during insertion. These simulation results are not in close agreement with the proposed formulas for secondary clustering.

	$m=101$			
n	C_n	L_n	C'_n	F_n
51	1.4126 ± 0.0025	5.000 ± 0.023	2.1182 ± 0.0030	$0. \pm 0.00017$
81	2.0605 ± 0.0044	11.552 ± 0.055	5.1282 ± 0.0085	$0. \pm 0.00017$
91	2.5654 ± 0.0052	18.510 ± 0.079	9.522 ± 0.013	0.00033 ± 0.00023
96	3.0358 ± 0.0069	26.64 ± 0.10	16.913 ± 0.021	0.0253 ± 0.0018
100	3.613 ± 0.011	36.70 ± 0.11	35.248 ± 0.034	0.5123 ± 0.0090

Table 3.10: Simulation results for Quadratic hashing

	$m=4999$			
n	C_n	L_n	C'_n	F_n
2499	1.4270 ± 0.0037	10.64 ± 0.31	2.1365 ± 0.0046	$0. \pm 0.020$
3999	2.1578 ± 0.0072	28.07 ± 0.90	5.620 ± 0.014	$0. \pm 0.019$
4499	2.781 ± 0.011	52.4 ± 1.8	11.193 ± 0.029	$0. \pm 0.018$
4749	3.435 ± 0.016	91.5 ± 2.5	21.650 ± 0.057	$0. \pm 0.015$
4949	4.987 ± 0.025	317.6 ± 9.7	99.262 ± 0.180	$0. \pm 0.0090$

Table 3.11: Simulation results for Quadratic hashing

References:
[Ackerman, 74], [Batagelj, 75], [Bell, 70], [Burkhard, 75], [Ecker, 74], [Knuth, 73], [Maurer, 68], [Radke, 70], [Samson *et al.*, 78], [Santoro, 76], [Wirth, 76].

3.3.6 Ordered hashing

It is easy to verify that the average access time for Uniform probing, Double hashing and Quadratic hashing not only depends on the keys, but also on the order in which these keys are inserted. Although the absolute order of insertion is difficult to alter, the algorithms described in this and the following sections will simulate altering the order of insertion. That is, if convenient, keys already in the table are moved to make room for newly inserted keys.

Ordered hashing is a composition of a hashing step, followed by Double hashing collision resolution. Furthermore, Ordered hashing reorders keys to simulate the effect of having inserted all the keys in increasing order. To achieve this effect, during insertion, smaller-value keys will cause relocation of larger-value keys found in their paths.

For the analysis of Ordered hashing we assume, as for Uniform probing, that the hashing function produces probing sequences without clustering. Let x be the probability that a randomly selected key in the file is less than the searched key. Then

$$\alpha = n/m$$

$$Pr\{A'_n(x) > k\} = \frac{n^{\underline{k}}}{m^{\underline{k}}} x^k$$

$$E[A_n'(x)] = \sum_{k \ge 0} \frac{n^{\underline{k}}}{m^{\underline{k}}} x^k = \frac{1}{1-\alpha x} - \frac{\alpha(1-\alpha)x^2}{(1-\alpha x)^3 m} + O(n^{-2})$$

$$C_n' = E[A_n'] = \sum_{k=0}^{n} \frac{n^{\underline{k}}}{m^{\underline{k}}(k+1)}$$

$$= -\alpha^{-1}\ln(1-\alpha) - \frac{1}{n}\left(\frac{3\alpha-2}{2(1-\alpha)} - \frac{1-\alpha}{\alpha}\ln(1-\alpha)\right) + O(n^{-2})$$

$$C_m' = E[A_m'] = H_{m+1}$$

The values for A_n and C_n are the same as those for Double hashing.

Ordered hashing: search

```
i := hashfunction( key ) ;
inc := increment( key ) ;
last := (i+(n-1)*inc) mod m;
while (i≠last) and (not empty(r[i])) and (r[i].k<key) do
        i := (i+inc) mod m;
if r[i].k=key then found(r[i])
              else notfound(key);
```

Ordered hashing: insertion

```
if n≥m then Error {*** table is full ***}
else begin
        i := hashfunction( key ) ;
        while (not empty(r[i])) and (not deleted(r[i]))
                and (r[i].k≠key) do begin
                if r[i].k > key then begin
                        {*** Exchange key and continue ***}
                        temp := key; key := r[i].k;      r[i].k := temp
                        end;
                i := (i+increment(key)) mod m
                end;
        if empty(r[i]) or deleted(r[i]) then begin
                {*** do insertion ***}
                r[i].k := key;
                n := n+1
                end
        else    Error {*** key already in table ***}
        end;
```

This variation of Double hashing reduces the complexity of the unsuccessful search to roughly that of the successful search at a small cost during insertion.

Table 3.12 shows simulation results for ordered hashing. We present the values for C_n' since the values for C_n and L_n are expected to be the same as those for Double hashing.

	C_n'		
n	$m=101$	n	$m=4999$
51	1.38634±0.00080	2500	1.38625±0.00082
81	1.9856±0.0023	3999	2.0115±0.0025
91	2.4871±0.0039	4499	2.5535±0.0043
96	2.9928±0.0059	4749	3.1540±0.0070
100	3.958±0.011	4949	4.640±0.018

Table 3.12: Simulation results on unsuccessful searches for Ordered hashing

References:
[Amble *et al.*, 74].

3.3.7 Reorganization scheme for Uniform probing: Brent's algorithm

Brent's reorganization scheme is based on Double hashing. This scheme will place a new key by moving forward at most one other key. The placement is done such that the total number of accesses (new key and old keys moved forward) is minimized. This is achieved by searching for the first empty location in the probing path of the new key or the paths of any of the keys in the path of the new key.

If $\alpha=n/m$ is the load factor, then

$$C_n \approx 1+\frac{\alpha}{2}+\frac{\alpha^3}{4}+\frac{\alpha^4}{15}-\frac{\alpha^5}{18}+\frac{2\alpha^6}{15}+\frac{9\alpha^7}{80}-\frac{293\alpha^8}{5670}-\frac{319\alpha^9}{5600}+\cdots$$

$$C_m \approx 2.4941...$$

α	C_α
0.5	1.2865
0.8	1.5994
0.9	1.8023
0.95	1.9724
0.99	2.2421
1.0	2.4941

Table 3.13: Exact values for C_α

It has been conjectured and verified by simulation that

$$L_m = O(\sqrt{m})$$

The values for the unsuccessful search are identical to those for Double hashing.

Brent's reorganization hashing: insertion

```
init := hashfunction( key );
inc  := increment( key );
for i:=0 to n do
        for j:=i downto 0 do begin
                jj := (init + inc*j) mod m;
                ii := (jj + increment(r[jj].k) * (i-j)) mod m;
                if empty(r[ii]) or deleted(r[ii]) then begin
                        {*** move record forward ***}
                        r[ii] := r[jj];
                        {*** insert new in r[jj] ***}
                        r[jj].k := key;
                        n := n+1;
                        goto 999  {*** return ***}
                        end
                end;
Error {*** table full ***};
999:
```

The above algorithm will not detect the insertion of duplicates, i.e. elements already present in the table.

The searching algorithm is identical to Double hashing.

This method improves the successful search at the cost of additional work during the insertion of new keys. This reorganization scheme allows us to completely fill a hashing table, while still keeping the average number of accesses bounded by 2.5. The length of the longest probe sequence, which is the actual worst case for a random file, is also significantly reduced.

For a stable table where its elements will be searched several times after insertion, this reorganization will prove very efficient.

Tables 3.14 and 3.15 summarize simulation results for Brent's reorganization scheme. The columns headed by I_n count the number of elements accessed to insert a new key in the table. I_n gives an accurate idea of the cost of the reorganization. Note that the variance on the number of accesses is also greatly reduced. The simulation results are in excellent agreement with the predicted theoretical results.

References:

[Brent, 73], [Feldman *et al.*, 73], [Knuth, 73].

$m = 101$				
n	C_n	$\sigma^2(A_n)$	L_n	I_n
51	1.2896 ± 0.0024	0.3084 ± 0.0033	3.099 ± 0.019	1.5154 ± 0.0056
81	1.6234 ± 0.0033	0.9038 ± 0.0074	5.113 ± 0.030	2.688 ± 0.013
91	1.8385 ± 0.0040	1.500 ± 0.013	6.669 ± 0.043	3.852 ± 0.022
96	2.0071 ± 0.0046	2.154 ± 0.020	8.223 ± 0.058	5.103 ± 0.034
100	2.2468 ± 0.0059	3.664 ± 0.042	11.460 ± 0.096	7.848 ± 0.071
101	2.3632 ± 0.0061	4.915 ± 0.061	14.03 ± 0.12	9.99 ± 0.10

Table 3.14: Simulation results for Brent's hashing

$m = 4999$				
n	C_n	$\sigma^2(A_n)$	L_n	I_n
2499	1.28650 ± 0.00067	0.2914 ± 0.0009	4.519 ± 0.036	1.4967 ± 0.0015
3999	1.60050 ± 0.00088	0.8074 ± 0.0020	7.780 ± 0.058	2.5544 ± 0.0035
4499	1.8052 ± 0.0011	1.3582 ± 0.0035	10.595 ± 0.079	3.6461 ± 0.0060
4749	1.9757 ± 0.0013	2.0420 ± 0.0057	13.89 ± 0.12	4.9425 ± 0.0098
4949	2.2451 ± 0.0018	3.993 ± 0.015	24.07 ± 0.23	8.416 ± 0.026
4999	2.4716 ± 0.0028	10.18 ± 0.14	84.9 ± 1.8	18.38 ± 0.23

Table 3.15: Simulation results for Brent's hashing

3.3.8 Reorganization scheme for Uniform probing: Binary tree hashing

Binary tree hashing is based on Double hashing. This scheme will insert a new key in the table by moving forward, if necessary, other keys. The placement is done such that the total number of accesses (new key and old keys moved forward) is minimized. This is achieved by searching for empty locations in the probing path of the new key or the paths of the keys in its path or the paths of any keys in the path of the path, and so on. The name "Binary tree" comes from the fact that the algorithm probes locations following a binary tree pattern.

If $\alpha = n/m$ is the load factor, then

$$C_n \approx 1 + \frac{\alpha}{2} + \frac{\alpha^3}{4} + \frac{\alpha^4}{15} - \frac{\alpha^5}{18} + \frac{2\alpha^6}{105} + \frac{83\alpha^7}{720} + \frac{613\alpha^8}{5760} - \frac{69\alpha^9}{1120} + \cdots$$

$$C_m \approx 2.13414...$$

If M_n is the number of keys that are moved forward for an insertion, then

$$M_n \approx \frac{\alpha^2}{3} - \frac{\alpha^3}{4} + \frac{2\alpha^4}{15} + \frac{\alpha^5}{9} + \frac{8\alpha^6}{105} - \frac{101\alpha^7}{720} - \frac{506\alpha^8}{2835} - \cdots$$

$$M_m \approx 0.38521...$$

It is conjectured, and supported by simulation, that

$$L_m = \log_2 m + O(1)$$

α	C_α	M_α
0.5	1.28517	0.06379
0.8	1.57886	0.17255
0.9	1.75084	0.24042
0.95	1.88038	0.29200
0.99	2.04938	0.35819
1.0	2.13414	0.38521

Table 3.16: Exact values for comparisons and moves

Binary tree reorganization hashing: insertion

```
init := hashfunction( key );
inc  := increment( key );
i := 0;        j := -1;
while (i≤n) and (j≤0) do begin
        j := SearchMove( init, inc, i );
        i := i+1
        end;
if j>-1 then begin
        {*** A hole was found, insert key ***}
        r[j].k := key;
        n := n+1
        end
else    Error {*** table is full ***};
```

Binary tree reorganization hashing: movement of entries

```
function SearchMove ( init, inc, level : integer ) : integer;
{*** Find the first hole (empty location) at the given depth
        in the binary tree spanned by a key ***}
label  999;
var    i, inc1, j, k : integer;
begin
i := (init + inc*level) mod m;
if empty(r[i]) or deleted(r[i]) then SearchMove := i
else   begin
       for j:=level-1 downto 0 do begin
               i := (init + inc*j) mod m;
               inc1 := increment( r[i].k );
               k := SearchMove( (i+inc1) mod m, inc1, level-j-1 );
               if k>-1 then begin
                       {*** A hole was found, move forward ***}
                       r[k] := r[i];
```

```
                SearchMove := i;
                goto 999 {*** return ***}
                end
            end;
        {*** Could not find hole ***}
        SearchMove := -1;
        end;
999:
end;
```

The above algorithm will not detect the insertion of duplicates, i.e. elements already present in the table.

This reorganization scheme significantly reduces the number of accesses for a successful search at the cost of some additional effort during the insertion of new keys. This algorithm is very suitable for building static tables, which will be searched often.

Table 3.17 summarizes simulation results for the Binary tree hashing reorganization scheme. The column headed by I_n counts the average number of elements accessed to insert a new key in the table. I_n gives an accurate idea of the cost of the reorganization. Note that the expected length of the longest probe sequence (L_n) is very small. On the other hand, the cost of inserting new elements is particularly high for full or nearly full tables. The simulation results are in excellent agreement with the predicted theoretical results.

$m=1999$				
n	C_n	L_n	I_n	M_n
1000	1.2834±0.0018	4.026±0.050	1.495±0.004	0.0632±0.0008
1599	1.5794±0.0022	6.143±0.086	2.629±0.010	0.1711±0.0011
1799	1.7536±0.0025	7.514±0.093	3.973±0.020	0.2353±0.0013
1899	1.8851±0.0022	8.564±0.089	5.982±0.032	0.2849±0.0012
1979	2.0615±0.0038	10.34±0.16	15.48±0.25	0.3463±0.0020
1999	2.1450±0.0039	12.30±0.22	231±40	0.3741±0.0021

Table 3.17: Simulation results for Binary tree hashing

References:
[Gonnet *et al.*, 77], [Gonnet *et al.*, 79], [Lyon, 79], [Maddison, 80], [Mallach, 77], [Rivest, 78].

3.3.9 Optimal hashing

Optimal hashing based on Double hashing or Uniform probing is the logical conclusion of the previous reorganization algorithms. Two complexity measures can be minimized: the average number of probes (C_n^{Opt}), or the number of accesses in the longest probe sequence (L_n^{Opt}).

The insertion algorithm is translated into an assignment problem (assignment of keys into table locations) and the cost of each assignment of K_i to location j is the number of probes necessary to locate the key K_i into location j.

For the minimax arrangement for Random probing we have

$$\ln m + \gamma + \frac{1}{2} + o(1) \leq E[L_m] = O(\ln m)$$

For the minimax arrangement for Uniform probing, we have the lower bound

$$\sum_{i=0}^{m-1} (-1)^{m-i-1} \binom{m}{i} \sum_{k \geq 0} \left(\frac{i^{\underline{k}}}{m^{\underline{k}}} \right)^m \leq E[L_m]$$

$$\lceil -\alpha^{-1} \ln (1-\alpha) \rceil \leq E[L_n]$$

For the minimum-average arrangement for Random probing and for Uniform probing we have:

$$1.688382... \leq C_m$$

These optimal algorithms are mostly of theoretical interest. The algorithms to produce these optimal arrangements may require $O(m)$ additional space during the insertion of new elements.

The following tables show some simulation results on optimal arrangements.

n	m	α	C_n^{Opt}	L_n
798	997	80%	1.4890 ± 0.0041	4.40 ± 0.11
897	997	90%	1.6104 ± 0.0043	5.147 ± 0.089
947	997	95%	1.6892 ± 0.0059	5.68 ± 0.12
987	997	99%	1.7851 ± 0.0058	6.77 ± 0.13
19	19	100%	1.729 ± 0.011	4.385 ± 0.071
41	41	100%	1.783 ± 0.011	5.29 ± 0.11
101	101	100%	1.798 ± 0.011	6.30 ± 0.18
499	499	100%	1.824 ± 0.011	7.92 ± 0.36
997	997	100%	1.8279 ± 0.0064	8.98 ± 0.38

Table 3.18: Simulation results for Optimal hashing (minave)

References:

[Gonnet *et al.*, 77], [Gonnet *et al.*, 79], [Gonnet, 77], [Gonnet, 81], [Lyon, 78], [Yao, 82].

3.3.10 Direct chaining hashing

This method makes use of both hashing functions and sequential lists (chains) in the following way. The hashing function first computes an index into the hashing table using the record key. This table location does not hold an actual record, but a pointer to a linked list of all records which hash to that

n	m	α	C_n	L_n^{Opt}
399	499	80%	1.4938 ± 0.0067	3.000 ± 0.030
449	499	90%	1.6483 ± 0.0079	3.050 ± 0.043
474	499	95%	1.6995 ± 0.0070	3.990 ± 0.020
494	499	99%	1.7882 ± 0.0077	5.120 ± 0.089
19	19	100%	1.749 ± 0.011	3.929 ± 0.062
41	41	100%	1.796 ± 0.010	4.665 ± 0.088
101	101	100%	1.807 ± 0.010	5.53 ± 0.14
499	499	100%	1.8300 ± 0.0081	7.38 ± 0.29

Table 3.19: Simulation results for Optimal hashing (minimax)

location. This is a composition of hashing with linked lists. The data structure used by this algorithm is described by $\{\mathbf{s}-\mathbf{D}\}_0^N$ where $\mathbf{s}-\mathbf{D}$ represents a linked list of data elements $\mathbf{D}$. Let P_n and P_n' be random variables which represent the number of pointers (chain links) inspected for the successful and unsuccessful searches respectively. Thus

$$P_n = A_n, \quad P_n' = A_n' + 1$$

The pertinent facts about this algorithm are listed below:

$$Pr\{chain\ with\ length\ i\} = \binom{n}{i} \frac{(m-1)^{n-i}}{m^n}$$

$$E[A_n] = C_n = 1 + \frac{n-1}{2m} \approx 1 + \frac{\alpha}{2}$$

$$\sigma^2(A_n) = \frac{(n-1)(n-5)}{12m^2} + \frac{n-1}{2m} \approx \frac{\alpha^2}{12} + \frac{\alpha}{2}$$

$$E[A_n'] = C_n' = \frac{n}{m}$$

$$\sigma^2(A_n') = \frac{n(m-1)}{m^2} \approx \alpha$$

$$E[L_n] = \Gamma^{-1}(m)\left(1 + \frac{\ln \alpha}{\ln \Gamma^{-1}(m)} + O\left(\frac{1}{\ln^2 \Gamma^{-1}(m)}\right)\right)$$

$$E[L_m] = \Gamma^{-1}(m) - \frac{3}{2} + \frac{\gamma - 1}{\ln \Gamma^{-1}(m)} + O\left(\frac{1}{\ln m}\right) + Q(\ln \ln m)$$

where $Q(x)$ is a periodic function of x and very small in magnitude.

Let S_r and S_p be the size of a record and the size of a pointer, then the expected storage used, $E[S_n]$, is

$$E[S_n] = (m+n)S_p + nS_r$$

Whenever $\frac{S_p}{S_r + S_p} < (1 - 1/m)^n \approx e^{-\alpha}$, this algorithm uses less storage than separate chaining hashing.

Descriptions of the search and insert algorithms are given below. For this algorithm, we will not use *r*, the array of records, but *ptrs* an array of heads of linked lists. The nodes of the linked list are the ones which contain the keys.

Direct chaining hashing: search

```
p = ptrs[ hashfunction( key ) ];
while ( p≠0 && key≠p→k )      p = p→next;
if ( p==0 ) notfound( key );
      else found( *p );
```

Direct chaining hashing: insertion

```
i = hashfunction( key );
ptrs[i] = newnode( key, ptrs[i] );
n++;
```

The above algorithm will not detect the insertion of duplicates, i.e. elements already present in the table.

The Direct chaining method has several advantages over open addressing schemes. It is very efficient in terms of the average number of accesses for both successful and unsuccessful searches, and in both cases the variance of the number of accesses is small. L_n grows very slowly with respect to n.

Unlike the case with open addressing schemes, contamination of the table due to deletions does not occur; to delete a record all that is required is an adjustment in the pointers of the linked list involved.

Another important advantage of Direct chaining is that the load factor α can be greater than 1; that is, we can have $n>m$. This makes the algorithm a good choice for dealing with files which may grow beyond expectations.

There are two slight drawbacks to the Direct chaining method. The first is that it requires additional storage for the $(m+n)$ pointers used in linking the lists of records. The second is that the method requires some kind of memory management capabilities to handle allocation and deallocation of list elements.

This method is very well suited for external search. In this case we will likely keep the array of pointers in main memory. Let E_n^b be the expected number of buckets accessed when Direct chaining hashing is used in external storage with bucket size b. Then

$$\alpha = n/m$$

$$E_n^b = \frac{n-1+m(b+1)}{2bm} + \frac{m(b^2-1)}{12bn} + \frac{m}{bn}\sum_{j=1}^{b-1} \frac{\omega_j}{(1-\omega_j)^2}\left(\frac{m+\omega_j-1}{m}\right)^n$$

$$= \frac{\alpha+b+1}{2b} + \frac{b^2-1}{12\alpha b} + O(k^{-\alpha}) \qquad (k>1)$$

where ω_j is the jth b^{th} root of unity.

$$E_n^2 = \frac{n-1}{4m} + \frac{3}{4} + \frac{m}{8n}(1-(1-2/m)^n)$$

References:
[Gonnet, 81], [Morris, 68], [Tai *et al.*, 80].

3.3.11 Separate chaining hashing

This method uses a hashing function to probe into an array of keys and pointers. Collisions are resolved by a linked list starting at the table entry.

The data structure used is described by $\{\mathbf{l-D}\}_0^N$ where

$$\mathbf{l-D : (D,[l-D]); (D,nil)} .$$

Let A_n and A_n' denote the number of accesses to records $\mathbf{l-D}$. The pertinent facts about this algorithm are

$$E[A_n'] = C_n' = \frac{n}{m} + (1-1/m)^n \approx \alpha + e^{-\alpha}$$

$$\sigma^2(A_n') = \frac{n(m-1)}{m^2} + \frac{m-1-2n}{m}(1-1/m)^n - (1-2/m)^n$$

$$\approx \alpha + (1-2\alpha)e^{-\alpha} - e^{-2\alpha}$$

The values for A_n, L_n and L_m coincide with those for Direct chaining hashing (3.3.10).

Let S_r and S_p be the size of a record and the size of a pointer, then the expected storage used, $E[S_n]$, is

$$E[S_n] = \left(n+m(1-1/m)^n\right)(S_r+S_p) \approx m(\alpha+e^{-\alpha})(S_r+S_p)$$

Whenever $\frac{S_p}{S_p+S_r} > (1-1/m)^n \approx e^{-\alpha}$ this algorithm uses less storage than Direct chaining hashing.

Descriptions of the search and insert algorithms are given below.

Separate chaining hashing: search

```
p = &r[ hashfunction( key) ];
while ( p≠0 && key≠p→k )     p = p→next;
if ( p==0 ) notfound( key );
       else  found( *p );
```

Separate chaining hashing: insertion

```
i = hashfunction( key );
if (empty(r[i])) /*** insert in main array ***/
                r[i].k = key;
        else    /*** insert in new node ***/
                r[i].next = newnode( key, r[i].next );
n++;
```

The above algorithm will not detect the insertion of duplicates, i.e. elements already present in the table.

This method has several advantages over open addressing schemes. It is very efficient in terms of the average number of accesses for both successful and unsuccessful searches, and in both cases the variance of the number of accesses is small. The length of the longest probe sequence, i.e. actual worst case, grows very slowly with respect to n.

Unlike open addressing schemes, contamination of the table due to deletions does not occur.

The load factor can go beyond 1 which makes the algorithm a good choice for tables that may grow unexpectedly.

This method requires some extra storage to allocate space for pointers. It also requires a storage allocation scheme to allocate and return space for records.

References:
[Bays, 73], [Johnson, 61], [Knuth, 73], [Larson, 82], [Morris, 68], [Quittner *et al.*, 81].

3.3.12 Coalesced hashing

Coalesced hashing is a hashing algorithm which resolves collisions by chaining. The chain pointers point to elements of the hashing array itself. The data structure used by this algorithm is described by $\{\mathbf{D,int}\}_0^N$ where the **int** is taken to be a "pointer" to the next element in the chain (an index into the array). The name "Coalesced hashing" comes from the fact that colliding records are stored in the main table, and keys hashing to other locations may share part of a chain.

The complexity measures for this algorithm are:

$$E[A_n] = C_n = 1 + \frac{1}{4}\left((1+2/m)^n - 1 - \frac{2n}{m}\right)$$

$$= 1+\frac{1}{4}\left(e^{2\alpha}-1-2\alpha\right)+O(m^{-1})$$

$$\sigma^2(A_n) = \frac{35}{144}-\frac{\alpha}{12}-\frac{\alpha^2}{4}+\frac{2\alpha-5}{8}e^{2\alpha}+\frac{4}{9}e^{3\alpha}-\frac{e^{4\alpha}}{16}+O(m^{-1})$$

$$E[A_n'] = C_n' = 1+\frac{m}{8n}\left((1+2/m)^n-1-\frac{2n}{m}\right)+\frac{n-1}{4m}$$

$$= 1+\frac{1}{8\alpha}(e^{2\alpha}-1-2\alpha)+\frac{\alpha}{4}+O(m^{-1})$$

$$\sigma^2(A_n') = -\frac{7}{36}+\frac{5\alpha}{24}-\frac{\alpha^2}{16}-\frac{37}{432\alpha}-\frac{1}{64\alpha^2}+$$
$$\left(-\frac{1}{16}-\frac{1}{16\alpha}+\frac{1}{32\alpha^2}\right)e^{2\alpha}+\frac{4e^{3\alpha}}{27\alpha}-\frac{e^{4\alpha}}{64\alpha^2}+O(m^{-1})$$

Descriptions of the search and insert algorithms are given below. The insertion algorithm uses the variable *nextfree* to avoid a repetitive search of the table for empty locations. This variable should be initialized to $m-1$ before starting to fill the table.

Coalesced hashing: search

```
i = hashfunction( key );
while ( i≠NULL && !empty(r[i]) && r[i].k≠key )    i = r[i].next;
if ( i==NULL || empty(r[i]) )   notfound( key );
            else   found( r[i] );
```

Coalesced hashing: insertion

```
i = hashfunction( key );
if ( empty(r[i]) )
        {
        r[i].k = key;
        r[i].next = NULL;
        n++;
        }
else    {/*** Find end of chain ***/
        while ( r[i].next≠NULL && r[i].k≠key )    i = r[i].next;
        if ( r[i].k==key )  Error /*** key already in table ***/;
        else {
             /*** Find next free location ***/
             while ( !empty(r[nextfree]) && nextfree≥0 )  nextfree--;
```

```
            if ( nextfree<0 ) Error /*** Table is full ***/;
            else {
                  r[i].next = nextfree;
                  r[nextfree].k = key;
                  r[nextfree].next = NULL;
                  n++;
                  }
            }
      }
```

Coalesced hashing is an efficient internal hashing algorithm at the cost of one integer (pointer to an array) per entry. Average and variance values for the successful and unsuccessful cases are very low.

This algorithm has some of the advantages of chaining methods without the need for dynamic storage allocation.

Due to the use of the variable *nextfree* the first collisions will fill the top of the table more quickly than the rest. This observation leads to a variation of the algorithm: *Coalesced hashing with cellar.* In this variation we leave the top part of the table (called the "cellar") to be filled only by collisions. The hashing function is now restricted to generate addresses in the lower part ("address region") of the table. The algorithms to perform searching and insertion with cellars are identical to the above; the only difference lies in the *hashfunction* which now generates integers over a restricted range.

Let us call m' the total size of the table (m "address" entries and $m'-m$ "cellar" entries). Let

$$\alpha = n/m$$

$$\beta = m/m'$$

and λ be defined as the positive solution of $e^{-\lambda}+\lambda = 1/\beta$. Then the complexity measures become:

$$E[A_n] = C_n = 1+\frac{\alpha}{2} \qquad \text{if } \alpha\le\lambda$$

$$= 1+\frac{1}{8\alpha}\left(e^{2(\alpha-\lambda)}\,1 -2(\alpha-\lambda)\right)(3-2/\beta+2\lambda) + \frac{\alpha+2\lambda-\lambda^2/\alpha}{4} + O\left(\frac{\log m'}{\sqrt{m'}}\right)$$

otherwise.

$$E[A_n'] = C_n' = \alpha+e^{-\alpha} \qquad \text{if } \alpha\le\lambda$$

$$= \frac{1}{\beta}+\frac{1}{4}\left(e^{2(\alpha-\lambda)}-1\right)(3-2/\beta+2\lambda) - \frac{\alpha-\lambda}{2} + O\left(\frac{\log m'}{\sqrt{m'}}\right)$$

otherwise.

For every value of α we could select an optimal value for β which minimizes either the successful or unsuccessful case. The value $\beta=0.853...$ minimizes the successful case for a full table while $\beta=0.782...$ does similarly for the unsuccessful case. The value $\beta=0.86$ appears to be a good compromise for both cases and a wide range of values for α.

References:
[Banerjee *et al.*, 75], [Bays, 73], [Chen *et al.*, to app.], [Knuth, 73], [Vitter, 80], [Vitter, 80], [Vitter, 81], [Vitter, 82], [Vitter, 82], [Vitter, 83], [Williams, 59].

3.3.13 Extendible hashing

Extendible hashing is a scheme which allows the hashing table to grow and guarantees a maximum of two external accesses per record retrieved. This scheme is best understood in terms of external storage searching. The structure is composed of a *directory* and *leaf-pages*

$$\textbf{directory : (N,}\{\textbf{[leaf}-\textbf{page]}\}_0^{2^N-1}\textbf{)}.$$

$$\textbf{leaf}-\textbf{page : (int,}\{\textbf{KEY}\}_1^b\textbf{)}.$$

where the directory consists of a set of pointers to leaf pages and the leaf-pages are buckets of size b with an additional depth indicator. Both directory and leaf depth indicators show how many bits of the hashing address are being considered; i.e. at depth d the hashing function returns a value in $0...2^d-1$. The depth of the leaf-pages is always less than or equal to the depth of the directory. Several directory entries may point to the same leaf page.

Basically this algorithm uses a composition of a hashing step with Sequential search in buckets. Every key resides in the bucket pointed by the directory entry indexed by the hashing value of the key. Collisions (overflow in the buckets) are handled by duplicating the directory.

Let $D_b(n)$ be the expected depth of the directory and m_d the number of entries in the directory and m_b the number of leaf-pages (buckets). We will assume that the number of keys is random, poisson-distributed with expected value n, then

$$d(n) = (1+1/b)\log_2 n$$

$$D_{(n)} = \sum_{k\geq 0}\left\{1-\left(\sum_{j=1}^{b} e^{-n2^{-k}}(n2^{-k})^j/j!\right)^{2^k}\right\}$$

$$= d(n)+\frac{\gamma+\ln((b+1)!)}{b\ln 2}+Q_1(d(n))+o(1)$$

$$E[m_d] = 1+\sum_{k\geq 0}2^k\left\{1-\left(\sum_{j=1}^{b} e^{-n2^{-k}}(n2^{-k})^j/j!\right)^{2^k}\right\}$$

$$= \left(\frac{\Gamma(1-1/b)}{\ln 2((b+1)!)^{1/b}} + Q_2(d(n)) \right) n^{1+1/b}(1+o(1))$$

$$\approx \frac{3.92}{b} n^{1+1/b}$$

and

$$E[m_b] = 1 + \sum_{k \geq 0} 2^k \sum_{j>b} e^{-n2^{-k}} (n2^{-k})^j / j!$$

$$= \frac{n}{b} \left(\frac{1}{\ln 2} + Q_3(\log_2 n) \right) + O(n^{1/2} \log n)$$

The functions $Q_i(x)$ are complicated periodic functions with period 1 and average value 0 (i.e. $\int_0^1 Q(x)\,dx = 0$).

Extendible hashing search

```
i := hashfunction( key ) mod m_d;
read-directory-entry( i ) into npage;
read-leaf-page( npage ) into r;
i := 1;
while (i<b) and (r[i].k ≠ key) do i := i+1;
if r[i].k = key then found( r[i] )
        else         notfound( key );
```

The insertions are straightforward if the corresponding bucket is not full. When an insertion causes a bucket to overflow, this bucket will be split into two and its depth increased by one. All the records in the bucket are then re-hashed with the new depth. Some of the pointers in the directory pointing to the splitting bucket may have to be redirected to the new bucket. If the depth of the bucket will exceed the depth of the directory, the directory is duplicated and its depth increased by one. Duplicating the directory implies a simple copy of its contents; no buckets are split. Certainly most buckets will be pointed to by two or more directory entries after the directory duplicates.

Assuming the existence of a fixed hash function $h_I(K)$, which returns an integer in a sufficiently large interval, the hashing function for level d can be implemented as

$$(h_I(K) \bmod 2^d)$$

This method allows graceful growth and shrinkage of an external hashing table. Assuming that the directory cannot be kept in internal storage, this method guarantees access to any record in two external accesses. This makes it a very good choice for organizing external files.

In case the directory can be kept in main memory, we can access records with a single external access, which is optimal.

The directory is $O(b^{-1}n^{1+1/b})$ in size. This means that for very large n or for relatively small bucket sizes, the directory may become too large. It is not likely that such a directory can be stored in main memory.

Insertions may be direct or may require the splitting of a leaf-page or may even require the duplication of the directory. This gives a bad worst-case complexity for insertion of new records.

Deletions can be done easily by marking or even by "folding" split buckets. Shrinking of the directory, on the other hand, is very expensive and may require $O(n)$ overhead to every deletion is some cases.

Below we give numerical values for several measures in Extendible hashing with poisson-distributed keys, for two different bucket sizes.

	$b=10$			$b=50$		
n	$D_b(n)$	$E[m_d]$	$E[m_b]$	$D_b(n)$	$E[m_d]$	$E[m_b]$
100	4.60040	25.8177	14.4954	1.71109	3.42221	2.92498
1000	8.45970	374.563	144.022	5.02284	32.7309	31.0519
10000	12.1860	4860.14	1438.01	8.99995	511.988	265.644
100000	16.0418	68281.7	14492.6	12.0072	4125.43	2860.62

Table 3.20: Exact values for Extendible hashing

References:
[Burkhard, 83], [Fagin *et al.*, 79], [Flajolet, 81], [King, 81], [Lloyd *et al.*, 82], [Lomet, 83], [Tamminen, 81], [Tamminen, 82], [Yao, 80].

3.3.14 Linear hashing

Linear hashing is a scheme which allows the hashing table to grow or shrink as records are inserted or deleted. This growth or shrinkage is continuous, one entry at a time, as opposed to Extendible hashing where the directory may duplicate due to a single insertion.

This scheme is best understood in terms of external representations. An external *bucket* is a physical record, convenient for performing input/output operations which may contain up to b records.

linear − hash − file : $\{$**bucket**$\}_0^{m-1}$.

bucket : ($\{$**KEY**$\}_1^b$,[**overflow**]).

overflow : ($\{$**KEY**$\}_1^{b_0}$,[**overflow**]) ; **nil** .

A bucket may receive more than b records, in which case the excess records are placed in a list of *overflow* buckets.

A file undergoes a *full expansion* when it grows from m_0 to $2m_0$. The process of growing from $2m_0$ to $4m_0$, etc. is an exact replica of the first full expansion.

This algorithm requires a *control function* $m = g(z)$, which regulates the growth of the array based on the number of records per storage used. We will use a control function that guarantees a constant storage utilization. For the storage utilization we will also consider the overflow buckets, i.e. we want to guarantee that $\frac{n}{b(m+m_{ov})} = \alpha$ or $m = \frac{n}{b\alpha} - m_{ov}$, where m_{ov} is the number of overflow buckets.

The following formulas indicate the limiting behaviour of Linear hashing, i.e. the limit when $n,m \rightarrow \infty$ while n/m remains constant.

$$C_n = \frac{1}{z_0} \int_{z_0}^{2z_0} \left(g(z)s(z/2) + (1-g(z))s(z) \right) dz$$

$$C_n' = \frac{1}{z_0} \int_{z_0}^{2z_0} \left(g(z)u(z/2) + (1-g(z))u(z) \right) dz$$

$$E[m_{ov}] = \frac{1}{6bz_0^2} \int_{z_0}^{2z_0} \left(2g(z)t(z/2) + (1-g(z))t(z) \right) dz$$

where

$$s(z) = 1 + \frac{1}{zb} \sum_{k \ge 0} (k+1) \sum_{j=1}^{b_o} \left(\frac{kb_o}{2} + j \right) P(b+kb_o+j,z)$$

is the expected number of accesses for a successful search in a single chain with relative load z and similarly

$$u(z) = 1 + \sum_{k \ge 0} (k+1) \sum_{j=1}^{b_o} P(b+kb_o+j,z)$$

for the unsuccessful search and

$$t(z) = b_o \sum_{k \ge 0} (k+1) \sum_{j=1}^{b_o} P(b+kb_o+j,z) = b_o(u(z)-1)$$

is the expected space taken by the overflow records of a single chain. $P(i,z)$ is a poisson probability distribution with parameter zb:

$$P(i,z) = \frac{e^{-zb}(zb)^i}{i!}$$

and finally $g(z)$ is the control function resulting from the policy of constant total storage utilization:

$$g(z) = \frac{zb/\alpha - t(z) - b}{2t(z/2) - t(z) + b}; \qquad z_0 = g^{-1}(0)$$

The hashing function for this algorithm depends on m and on m_0 as well as on the key. Each time that m increases by one, the hashing function changes, but this change is minimal in the following sense:

$$h_{m+1}(K) = \begin{cases} h_m(K) & \textit{iff } h_m(K) \neq m - m_0 \\ m \textit{ or } m - m_0 & \text{otherwise} \end{cases}$$

A common implementation of this function, assuming that we have available a basic hash function $h_I(K)$ which transforms the key into an integer in a sufficiently large interval, is:

Hashing function for Linear hashing

$i := h_I($ *key* $);$
if $(i$ **mod** $m_0) < m - m_0$ **then** *hashfunction* $:= i$ **mod** $(2m_0);$
else *hashfunction* $:= i$ **mod** $m_0;$

The algorithm, as described, suffers from a discontinuity. At the beginning of a full expansion most chains are of the same expected length, while at the end of an expansion, the last chains to be expanded will have an expected length about twice the average. This problem may be remedied by splitting the full expansion into partial expansions. For example, we can expand first from m_0 to $3m_0/2$ using the entries from 0 to $m_0/2-1$ and from $m_0/2$ to m_0-1 and secondly from $3m_0/2$ to $2m_0$ using the entries from $0...m_0/2-1$, $m_0/2...m_0-1$ and $m_0...3m_0/2-1$. By doing partial expansions the discontinuities are much less pronounced.

Linear hashing is a descendant of Virtual hashing. In Virtual hashing the main difference is that the file is duplicated in size in a single step, when appropriate, rather than entry by entry.

These methods are supposed to be excellent methods for storing large tables which require quick access in external storage.

References:
[Larson, 80], [Larson, 82], [Larson, 82], [Litwin, 78], [Litwin, 80], [Lloyd *et al.*, 82], [Mullin, 81], [Ramamohanarao *et al.*, 82].

3.3.15 External hashing using minimal internal storage

These algorithms are designed to work for external files. Under this assumption, all internal computations are viewed as insignificant when compared to an external access to the file. The goal is to minimize the number of external accesses, at the cost of maintaining some additional "indexing" information in main memory.

The algorithms described in this section act as "filters" on the external accesses of most other hashing algorithms (Uniform probing, Random probing, Double hashing, etc.). In other words, the searching is conducted as in the basic hashing algorithms, except that instead of accessing the external table directly, we first "consult" the internal information. When an access to external storage is allowed, it is either guaranteed to succeed, or has a very high probability of succeeding.

These algorithms will use the *signature* of a key. A signature function is a hashing function that returns a sequence of bits. It can be viewed as returning a uniformly distributed real number in [0,1) and the sequence of bits is given by its binary representation.

The minimization of resources can be approached in two distinct ways:

(a) guarantee exactly one external access (optimal) for each key while minimizing the additional internal storage required; or

(b) given a fixed amount of internal storage, minimize the number of external accesses.

Let us call *k-prefix* the first k bits of the signature of a key. To solve the first problem we will construct the following table. For each table location we will code the following information: (i) the location is empty or (ii) the location is occupied and the key stored in this location has a prefix of length k. The prefix stored is the shortest possible required to distinguish the stored key from all other keys that probe this location. Note that on the average, only $C_n - 1$ other keys probe an occupied location. This algorithm requires building a table of variable length prefixes, hence we will call it *Variable-length signatures.*

Let $m_b(n)$ be the average number of internal bits required by these algorithms; then if the main hashing algorithm is Uniform probing or Random probing we have the following lower and upper bounds (the upper bounds represent the behaviour of the above algorithm):

$$\alpha = n/m$$

$$\frac{n}{\ln 2}\left(\alpha + (1-\alpha)\ln(1-\alpha) - \int_0^\alpha \frac{x \ln x}{1-x}\,dx\right) + O(1) \le m_b(n) \le$$

$$\log_2(-\ln(1-\alpha)) + O(1)$$

$$\frac{m\pi^2}{6 \ln 2} - \frac{3 \log_2 m}{2} + O(1) \le m_b(m) \le \log_2 \log_2 m + O(1)$$

A better lower bound is obtained for *memoriless* hashing algorithms. Let us call an algorithm memoriless if it does not store any information gained from previous probes, except the implicit fact that they failed. All the hashing algorithms in this section are memoriless.

$$\log_2 \beta + \frac{1}{\ln 2}\left(\frac{1}{2\beta} + \frac{5}{12\beta^2} + O(\beta^{-3})\right) + O\left(\frac{1}{m-n}\right) \le m_b(n)$$

where $\beta = -\ln(1-\alpha)$ and

$$\log_2 H_m + O\left(\frac{1}{\ln m}\right) \le m_b(m)$$

For the second problem we now restrict ourselves to using a fixed, small number, d, of bits per location. The goal is now to reduce the number of external accesses. If we store in each location the *d-prefix* of the stored key, we reduce the unnecessary accesses by a factor of 2^d. For this algorithm

$$C_n^d = 1-2^{-d}(\ln(1-\alpha)/\alpha+1)+O(\frac{1}{m-n})$$

This algorithm can be extended for external buckets containing b records each. For this extension, we cannot keep a signature of all the records in the bucket, but instead we keep a *separator*. A separator is a prefix long enough to distinguish between the signatures of the records which are stored in the bucket (lower prefixes) and those of the records which overflowed to other buckets (larger prefixes). This algorithm may displace records with high prefixes as records with smaller prefixes are inserted.

Finally, by selecting a fixed length separator and by moving forward records that would force a larger separator, an optimal and very economical algorithm is obtained. In this last case, there is a limit on the load of the external file. In other words, an insertion may fail although there is still room in the table (this happens when all the buckets are full or their separators are fully utilized).

Although these algorithms require internal tables, the actual sizes for real situations are affordable by almost any system. The reduction in the number of external accesses is very attractive. These methods are more economical in internal storage than Extendible hashing with an internally stored directory.

References:
[Gonnet *et al.*, 82].

3.4 Recursive structures search

3.4.1 Binary tree search

The Binary tree search is an algorithm for searching a lexicographically ordered binary tree. Without loss of generality we may assume that the left descendant nodes of any node contain keys whose values are less than or equal to the root, and that the right descendant nodes contain keys whose values are greater than the root.

Let A_n be the number of accesses (or node inspections) made in the course of a successful search for a given key in a binary tree of size n, and let A_n' be the number of accesses made in the course of an unsuccessful search of size n.

The symbol $h(n)$ denotes the *height* of a tree of size n, i.e. the number of nodes in the longest path through the tree. With this definition, a null tree has height 0, a single node tree has height 1. The *depth* of a node in a tree is the distance from the root to that node; thus the depth of the root is 0.

Several variations on the basic binary tree structure arise with the introduction of semantic rules or constraints such as height balance, weight balance, or heuristic organization schemes. The Pascal data definition and search algorithm for all binary trees are given below.

Tree definition

```
type   tree = ↑node;
       node =        record
              k : typekey;
              left, right : tree
              end;
```

Binary tree search

```
procedure search( key : typekey; t : tree );
begin
if t=nil then {*** Not Found ***}
              notfound( key )
else if t↑.k = key then
              {*** Found ***}
              found( t↑ )
else if t↑.k < key then      search( key, t↑.right )
        else                 search( key, t↑.left )
end;
```

The number of different trees with n nodes is

$$t_n = \frac{1}{n+1}\binom{2n}{n}$$

and the associated generating function is:

$$T(z) = \sum_{n \geq 0} t_n z^n = 1 + zT^2(z) = \frac{1-\sqrt{1-4z}}{2z}$$

The internal path length, I_n, of a tree with n nodes is defined as the sum of the depths of all its nodes. The external path length, E_n, of a tree is the sum of the depths of all its external nodes. For any binary tree

$$E_n = I_n + 2n$$

Let a_k be the expected number of nodes at depth k and let b_k be the expected number of external nodes at depth k in a binary tree with n nodes. Then we have the associated generating functions

$$A(z) = \sum_k a_k z^k$$

$$B(z) = \sum_k b_k z^k = (2z-1)A(z)+1$$

and

$$A(1) = B(1)-1 = n$$
$$A'(1) = E[I_n]$$
$$B'(1) = E[E_n]$$

For a successful search we have

$$C_n = E[A_n] = \frac{E[I_n]}{n}+1 = \frac{A'(1)}{n}+1 = (1+1/n)C_n'-1$$
$$\sigma^2(A_n) = 3C_n-2+\frac{A''(1)}{n}-C_n^2$$
$$1 \leq A_n \leq h(n)$$

and for an unsuccessful search

$$C_n' = E[A_n'] = \frac{E[E_n]}{n+1} = \frac{B'(1)}{n+1}$$
$$\sigma^2(C_n') = \frac{B''(1)}{n+1}+C_n'(1-C_n')$$
$$1 \leq A_n' \leq h(n)$$

The ordered binary tree is a structure which allows us to perform many operations efficiently: inserting takes a time of $O(h(n))$; deleting a record also takes $O(h(n))$; finding the maximum or minimum key requires $O(h(n))$ comparisons; and retrieving all the elements in ascending or descending order can be done in a time of $O(n)$. With small changes, it permits the retrieval of the kth ordered record in the tree in $O(h(n))$.

References:
[Aho *et al.*, 74], [Aho *et al.*, 83], [Burge, 76], [Dasarathy *et al.*, 80], [Driscoll *et al.*, 78], [Flajolet *et al.*, 78], [Flajolet *et al.*, 79], [Flajolet *et al.*, 79], [Flajolet *et al.*, 80], [Gill, 80], [Gotlieb *et al.*, 78], [Horowitz *et al.*, 76], [Kleitman *et al.*, 78], [Knott, 77], [Knuth, 73], [Lee, 80], [Munro *et al.*, 83], [Nievergelt, 74], [Ottmann *et al.*, 82], [Poblete, 82], [Proskurowski, 80], [Rotem *et al.*, 78], [Ruskey *et al.*, 77], [Solomon *et al.*, 80], [Standish, 80], [Stephenson, 80], [Strong *et al.*, 79], [vanLeeuwen *et al.*, 83], [Wirth, 76], [Yongjin *et al.*, 79].

3.4.1.1 Randomly generated binary trees

These structures are also known as Random search trees. Such trees are generated by taking elements in a random order and inserting them into an empty tree using the algorithm described below. Ordered binary search trees are normally considered to be created in this way. The efficiency measures for searching such trees are

$$C_n = 1 + n^{-1}\sum_{i=0}^{n-1} C_i'$$

$$E[A_n] = C_n = 2(1+1/n)H_n - 3 \approx 1.3863 \log_2 n - 1.8456$$

$$\sigma^2(A_n) = (2+10/n)H_n - 4(1+1/n)(H_n^2/n + H_n^{(2)}) + 4$$

$$\approx 1.3863 \log_2 n - 1.4253$$

$$E[A_n'] = C_n' = 2H_{n+1} - 2 \approx 1.3863 \log_2 n - 0.8456$$

$$\sigma^2(A_n') = 2H_{n+1} - 4H_{n+1}^{(2)} + 2 \approx 1.3863 \log_2 n - 3.4253$$

$$3.634...\ln n + o(\ln n) \leq E[h(n)] \leq 4.31107...\ln n - 2.80654...\ln \ln n + O(1)$$

where $H_n = \sum_{i=1}^{n} 1/i$ is the nth harmonic number, and $H_n^{(2)} = \sum_{i=1}^{n} i^{-2}$ is the nth bi-harmonic number.

Numerical evidence and simulation results indicate that the upper bound is tight.

Binary tree insertion

```
procedure insert( key : typekey; var t : tree );
begin
if t = nil then
                    t := newnode( key, nil, nil )
else if t↑.k = key then
                    Error {*** Key already in table ***}
else if t↑.k < key then        insert( key, t↑.right )
          else                 insert( key, t↑.left )
end;
```

At the cost of two extra pointers per element, randomly generated binary trees display excellent behaviour in searches. Unfortunately, the worst case can be generated when the elements are sorted before they are put into the tree. In particular, if any subset of the input records is sorted, it will cause the tree to degenerate badly. Compared to the Random binary trees of the next section, however, ordered binary trees generated from random input are exceptionally well behaved.

Table 3.21 gives numerical values for several efficiency measures in trees of various sizes.

References:

[Brinck *et al.*, 81], [Eppinger, 83], [Guibas, 75], [Itai *et al.*, 80], [Jonassen *et al.*, 78], [Knott, 82], [Knuth, 73], [Knuth, 74], [Knuth, 77], [Palmer *et al.*, 74], [Robson, 79], [Robson, 82], [Sprugnoli, 81], [Wilson, 76], [Wright, 81].

n	C_n	$\sigma^2(A_n)$	C_n'	$\sigma^2(A_n')$	$E[h(n)]$
5	2.4800	1.1029	2.900	0.9344	3.8000
10	3.4437	2.1932	4.0398	1.8076	5.6411
50	6.1784	5.6159	7.0376	4.5356	10.8103
100	7.4785	7.2010	8.3946	5.8542	13.2858
500	10.6128	10.7667	11.5896	9.0179	19.3359
1000	11.9859	12.2391	12.9729	10.3972	22.0362
5000	15.1927	15.5608	16.1894	13.6105	28.4283
10000	16.5772	16.9667	17.5754	14.9961	31.2216

Table 3.21: Exact complexity measures for Binary search trees

3.4.1.2 Random binary trees

When speaking of Random binary trees, we consider the situation where all possible trees with the same number of nodes are equally likely to occur. In this case,

$$E[A_n] = \frac{4^n - \frac{3n+1}{n+1}\binom{2n}{n}}{\frac{n}{n+1}\binom{2n}{n}} = \sqrt{\pi n}\left(1+\frac{9}{8n}+\frac{17}{128n^2}+O(n^{-3})\right)-3-\frac{1}{n}$$

$$E[A_n'] = \frac{4^n - \frac{n-1}{n+1}\binom{2n}{n}}{\binom{2n}{n}} = \sqrt{\pi n}\left(1+\frac{1}{8n}+\frac{1}{128n^2}+O(n^{-3})\right)-\frac{n-1}{n+1}$$

$$E[h(n)] = 2\sqrt{\pi n}+O(n^{1/4+\delta}) \quad (\textit{for any } \delta>0)$$

$$\sigma^2(I_n) = \left(\frac{10}{3}-\pi\right)n^3-\frac{n^2\sqrt{\pi n}}{2}+9(1-\pi/4)n^2-\frac{25n\sqrt{\pi n}}{16}+O(n)$$

If $t_{n,h}$ is the number of trees of height h and size n, then the associated generating function is

$$B_h(z) = \sum_{n=0}^{\infty} t_{n,h}z^n = zB_{h-1}^2(z)+1$$

When all trees of height h are considered equally likely to occur, then

$$E[nodes] = (0.62896...)2^h-1+O(\delta^{-2^h}) \quad (\delta>1)$$

This situation is primarily a theoretical model. In practice, very few situations give rise to random trees.

References:
[Flajolet *et al.*, 80], [Kemp, 79], [Kemp, 80], [Knuth, 73].

3.4.1.3 Height balanced trees

These are also known as AVL trees. Height balanced trees have the property that any two subtrees at a common node differ in height by 1 at the most. This balance property can be efficiently maintained by means of a counter in each node, indicating the difference in height between the right and left subtrees, $h(right)-h(left)$. The data structure used by an AVL tree is defined by **bt- (int,KEY)-LEAF.**

Because of the height-balancing, the total height of a tree with n nodes is bounded by

$$\lceil \log_2 n+1 \rceil \leq h(n) \leq 1.44042... \log_2 (n+2) - 0.32772...$$

There are AVL trees for which

$$C_n \geq 1.27937... \log_2 n + O(1)$$

Let R_n indicate the average number of rotations required during the insertion of the nth key into a random tree. Then

$$0.42531... \leq R_n \leq 0.72118...$$

Let B_n be the average number of AVL nodes that are completely height balanced. Then

$$0.59601...n+O(n^{-6}) \leq B_n \leq 0.73541...n+O(n^{-6})$$

Let $t_{n,h}$ be the number of height balanced trees of height h and size n. The associated generating function is

$$T_h(z) = \sum_{n \geq 0} t_{n,h} z^n = z T_{h-1}(z) \left(2T_{h-2}(z) + T_{h-1}(z) \right)$$

If we assume that all trees of height h are equally likely to occur, the average number of nodes in a balanced tree of height h is

$$E[nodes] = (0.70118...) 2^h$$

Below we give the description of the AVL insertion algorithm. The insertion algorithm uses an additional balance counter in each node of the tree, *bal*. The range of this balance field is $-2...2$. The procedures *rrot()* and *lrot()* which perform right and left rotations are common to several algorithms and are described in section 3.4.1.8.

Height balanced tree (AVL) insertion

```
function insert( key : typekey; var t : tree ) : integer;
var     incr : integer;
begin
insert := 0;
if t = nil then         begin
                t := newnode( key, nil, nil );
                t↑.bal := 0;
                insert := 1
                end
else if t↑.k = key then
                Error {*** Key already in table ***}
else    with t↑ do      begin
```

```
    if k < key then     incr := insert( key, right )
            else   incr := -insert( key, left );
    bal := bal + incr;
    if (incr ≠ 0) and (bal ≠ 0) then
            if bal < -1 then
                    {*** left subtree too tall: right rotation needed ***}
                    if left↑.bal < 0 thenrrot( t )
                            else   begin lrot( left ); rrot( t ) end
            else   if bal > 1 then
                    {*** right subtree too tall: left rotation needed ***}
                    if right↑.bal > 0 then     lrot( t )
                            else   begin rrot( right ); lrot( t ) end
            else   insert := 1;
    end
end;
```

AVL trees are of practical and theoretical importance as they allow searches, insertions and deletions in $O(\log n)$ time in the worst case.

The balance information in an AVL tree needs to represent 3 cases (5 cases for some particular implementations). This requires 2 (or 3) bits per node. It is not likely that this unit of storage is available, and a larger amount will be allocated for this purpose. Although a lot of emphasis has been placed on reducing the amount of extra storage used, the storage required by balance information is of little practical significance. If enough space is available it is best to store the height of the subtree, which contains more useful information and leads to simpler algorithms. Note that using 6 bits for height information we could store trees with up to 0.66×10^{13} nodes.

The constraint on the height balance can be strengthened to require that either both subtrees be of the same height or the right-side one be taller by one. These trees are called *One-sided height balanced, OSHB,* trees. In this case only one bit per node is required to store the balance information. Insertions in OSHBs become more complicated though; in particular, insertions in $O(\log n)$ time are extremely complicated.

Similarly, the constraint on the balance may be relaxed. One option is to allow the height of subtrees to differ at most by k. These trees are called *k-height balanced,* $HB[k]$, trees.

Table 3.22 shows some simulation results for AVL trees. C_n indicates the average number of comparisons required in a successful search, R_n is the average number of rotations (single or double) required by an insertion, and $E[h(n)]$ indicates the average height of a tree of size n.

The values for C_n' can be calculated from the above, e.g. for all binary trees $C_n' = (C_n+1)/(1+1/n)$.

Note that the variance on the number of accesses for the successful search is small, and appears to be bounded for increasing tree size.

n	C_n	$E[h(n)]$	R_n
5	2.2	3.0	0.21333
10	2.90712 ± 0.00010	4.0	0.31814 ± 0.00046
50	4.93035 ± 0.00037	6.9472 ± 0.0020	0.42683 ± 0.00056
100	5.88879 ± 0.00051	7.99905 ± 0.00055	0.44443 ± 0.00064
500	8.19213 ± 0.00079	10.9230 ± 0.0068	0.46131 ± 0.00055
1000	9.2017 ± 0.0012	11.9984 ± 0.0016	0.46357 ± 0.00061
5000	11.5552 ± 0.0025	14.936 ± 0.021	0.46541 ± 0.00060
10000	12.5676 ± 0.0030p	15.9960 ± 0.0078	0.46565 ± 0.00061

Table 3.22: Simulation results for AVL trees

From the above results we can see that the value for C_n is close to the value of $\log_2 n$; in particular under the arbitrary assumption that

$$C_n = \alpha \log_2 n + \beta$$

for $n \geq 500$, then

$$\alpha = 1.01207 \pm 0.00061; \qquad \beta = -0.8825 \pm 0.0058$$

References:
[Adel'son-Vel'skii *et al.*, 62], [Aho *et al.*, 74], [Baer *et al.*, 77], [Brown, 78], [Brown, 79], [Brown, 79], [Foster, 65], [Foster, 73], [Gonnet *et al.*, 83], [Guibas *et al.*, 78], [Hedetniemi, 80], [Hirschberg, 76], [Karlton *et al.*, 76], [Knuth, 73], [Kosaraju, 78], [Luccio *et al.*, 76], [Luccio *et al.*, 78], [Luccio *et al.*, 78], [Luccio *et al.*, 80], [Mehlhorn, 79], [Mehlhorn, 82], [Ottmann *et al.*, 78], [Ottmann *et al.*, 79], [Ottmann *et al.*, 79], [Ottmann *et al.*, 80], [Raiha *et al.*, 79], [Tan, 72], [Wright, 81], [Zaki *et al.*, 79], [Ziviani *et al.*, 82], [Zweben *et al.*, 78].

3.4.1.4 Weight balanced trees

These are also known as $BB(\alpha)$ trees. Weight balanced trees are Binary search trees which obey a balance criterion on the subtrees of every node. Each node of the tree has a weight attached to it. A tree is said to be of *weighted balance* α or of *bounded balance* α, or in the set $BB[\alpha]$, for $0 \leq \alpha \leq 1/2$, if every node in the tree has balance, $\rho(t)$, between α and $1-\alpha$. The balance of a node is defined as

$$\rho(t) = \frac{\textit{number of leaves in } \mathrm{t\uparrow.left}}{\textit{number of leaves in } \mathrm{t}}$$

The empty binary tree is in $BB[\alpha]$ by convention.

The set $BB[\alpha]$ becomes more restricted as α goes from 0 to 1/2. $BB[0]$ is the class of all Binary search trees, and $BB[1/2]$ is the class of completely balanced Binary search trees of $n = 2^h - 1$ nodes. Interesting $BB[\alpha]$ trees are the ones with $2/11 \leq \alpha \leq 1 - \sqrt{2}/2$. For these α, a balanced tree which is updated by an insertion or a deletion can be rebalanced with at most one rotation per level.

For any value of α,

$$\lceil \log_2(n+1) \rceil \le h(n) \le -\frac{\log_2 n}{\log_2(1-\alpha)} + O(1)$$

$$C_n \le -\frac{\log_2 n}{\alpha \ln \alpha + (1-\alpha)\ln(1-\alpha)} + O(1)$$

For any sequence of n updates (insertions and/or deletions), the worst-case average number of rotations is bounded by a constant which depends only on α:

$$R_n \le c(\alpha)$$

For the class of trees $BB[1-\sqrt{2}/2]$

$$\lceil \log_2(n+1) \rceil \le h(n) \le 2\log_2(n+1)$$

$$\le 2\log_2(n+3) - 2.44549...$$

$$C_n \le 1.14622... \log_2 n + O(1)$$

Let R_n be the average number of rotations per insertion in a $BB[1-\sqrt{2}/2]$ tree after the random insertion of n keys into the empty tree. Let $f(\beta)$ be the fraction of internal nodes with weight balance factor exactly β or $1-\beta$ in a random $BB[1-\sqrt{2}/2]$ tree with n keys. Then

$$R_n \ge 0.40238...$$

$$0.54291... \le f(1/2) \le 0.72593...$$

$$0.17231... \le f(1/3) \le 0.34801...$$

$$0.05405... \le f(2/5) \le 0.22975...$$

Below we give a description of the insertion algorithm for weight balanced trees with $\alpha = 1-\sqrt{2}/2 = 0.292893...$ The procedures *rrot()* and *lrot()*, which perform right and left rotations, are common to several algorithms and are described in section 3.4.1.8. The insertion algorithm uses a weight counter in each node of the tree, *weight*. For any node t, *t↑.weight* contains the number of external nodes in the subtree rooted at t. We use for convenience the function $wt(t)$ which returns 1 if the tree t is nil or *t↑.weight* otherwise.

Weight balanced tree insertion

```
procedure insert( key : typekey; var t : tree );
var    wbal : real;
begin
if t = nil then       begin
              t := newnode( key, nil, nil );
              t↑.weight := 2
              end
```

```
else if t↑.k = key then
              Error {*** Key already in table ***}
else    with t↑ do    begin
        if k < key then     insert( key, right )
              else   insert( key, left );
        weight := wt( left ) + wt( right );
        wbal := wt(left) / wt(t);
        if wbal > 0.707011 then
              {*** left subtree too heavy: right rotation needed ***}
              if wt(left↑.left) / wt(left) > 0.414213 then     rrot( t )
                    else   begin lrot( left ); rrot( t ) end
        else  if wbal < 0.292893 then
              {*** right subtree too heavy: left rotation needed ***}
              if wt(right↑.left) / wt(right) < 0.585786 then  lrot( t )
                    else   begin rrot( right ); lrot( t ) end
        end
end;
```

Although the insertion algorithm is coded using real arithmetic, this is not really needed. E.g. $\sqrt{2}/2$ can be approximated by its convergents 2/3, 5/7, 12/17, 29/41, 70/99, ... In case integer arithmetic must be used, the first test can be rewritten, for example, as

if 99*wt(t↑.left) > 70*wt(t) then ...

Table 3.23 shows some simulation results on weight balanced trees for $\alpha=1-\sqrt{2}/2$. C_n indicates the average number of comparisons required in a successful search, R_n is the average number of rotations (single or double) required by an insertion and $E[h(n)]$ indicates the average height of a tree of size n.

n	C_n	$E[h(n)]$	R_n
5	2.2	3.0	0.21333
10	2.9	4.0	0.32559±0.00062
50	4.94402±0.00051	7.0263±0.0031	0.40918±0.00071
100	5.90785±0.00068	8.2075±0.0065	0.42171±0.00071
500	8.2307±0.0015	11.261±0.016	0.43196±0.00073
1000	9.2453±0.0021	12.588±0.026	0.43263±0.00072
5000	11.6176±0.0045	15.653±0.059	0.43462±0.00076
10000	12.6498±0.0066	17.000±0.072	0.43442±0.00074

Table 3.23: Simulation results for weight balanced trees

From the above results we can see that the value for C_n is close to the value of $\log_2 n$, in particular under the arbitrary assumption that

$$C_n = \alpha \log_2 n + \beta$$

for $n \geq 500$, then

$$\alpha = 1.0203 \pm 0.0011; \qquad \beta = -0.919 \pm 0.011$$

References:
[Baer *et al.*, 77], [Blum *et al.*, 78], [Knuth, 73], [Nievergelt *et al.*, 73], [Unterauer, 79].

3.4.1.5 Balancing by internal path reduction

These are also known as *weight balanced* trees. These trees are similar to the trees described in the previous section, except that rotations are made only when they can reduce the total internal path of the tree. For this reason these trees are also known as *path trees*. In summary, a single left rotation is performed whenever

wt(t↑.left) < wt(t↑.right↑.right)

and a double left rotation when

wt(t↑.left) < wt(t↑.right↑.left)

and right rotations for the symmetric cases. For these balance conditions we have:

$$\lceil \log_2(n+1) \rceil \leq h(n) \leq 1.44042...\log_2 n - 0.32772...$$

$$C_n \leq \frac{5\log_3 2}{3}\log_2 n + O(1) = 1.05155...\log_2 n + O(1)$$

The amortized worst-case number of rotations per insertion is bounded by

$$R_n \leq 0.44042...\log_2 n + O(1)$$

The amortized worst-case number of rotations per deletion is bounded by

$$R_n \leq 0.42062...\log_2 n + O(1)$$

In the worst case, for a single insertion or deletion,

$$R_n = O(n)$$

Below we give a description of the insertion algorithm. The insertion code uses the procedure *checkrot* which checks the balancing, performs any necessary rotation and checks whether further rotations may be needed. The procedures *rrot()* and *lrot()*, which perform right and left rotations, are common to several algorithms and are described in section 3.4.1.8. For any node t, t↑*.weight* contains the number of external nodes in the subtree rooted at t. We use for convenience the function $wt(t)$ which returns 1 if the tree t is nil or t↑*.weight* otherwise.

Internal path reduction trees: insertion

```
procedure checkrots( var t : tree );
{*** check need for rotations ***}
var     wl, wr : integer;
begin
if t ≠ nil then with t↑ do begin
        wl := wt( left );
        wr := wt( right );
        if wr>wl then begin
                {*** left rotation needed ***}
                if wt( right↑.right ) > wl then
                        begin lrot( t ); checkrots( left ) end
                else if wt( right↑.left ) > wl then begin
                        rrot( right ); lrot( t );
                        checkrots( left );    checkrots( right )
                        end
                end
        else if wl>wr then begin
                {*** right rotation needed ***}
                if wt( left↑.left ) > wr then
                        begin rrot( t ); checkrots( right ) end
                else if wt( left↑.right ) > wr then begin
                        lrot( left );   rrot( t );
                        checkrots( left );    checkrots( right )
                        end
                end
        end
end;

procedure insert( key : typekey; var t : tree );
begin
if t = nil then         begin
                t := newnode( key, nil, nil );
                t↑.weight := 2
                end
else if t↑.k = key then
                Error {*** Key already in table ***}
else    with t↑ do      begin
        if k < key then         insert( key, right )
                else    insert( key, left );
        weight := wt( left ) + wt( right );
        checkrots( t )
        end
end;
```

Although these trees are in the class $BB(1/3)$, there are some important restrictions on the rotations. This makes their performance superior to the $BB(1/3)$ trees.

A natural extension of this algorithm is to perform rotations only when the difference in weights is k or larger. This extension is called *k-balancing*. For these trees the main complexity measures remain of the same order, while the number of rotations is expected to be reduced by a factor of k.

$$h^k(n) \leq 1.44042...\log_2(n-k+2)+k-1.32772...$$

$$C_n^k \leq 1.05155...\log_2 n + O(1)$$

Table 3.24 shows simulation results for these trees. C_n indicates the average number of comparisons required in a successful search, R_n is the average number of rotations (single or double) required by an insertion and $E[h(n)]$ indicates the average height of a tree of size n.

n	C_n	$E[h(n)]$	R_n
5	2.2	3.0	0.21333
10	2.9	4.0	0.33015±0.00062
50	4.90330±0.00027	6.9327±0.0028	0.47210±0.00082
100	5.85596±0.00036	8.0025±0.0014	0.49763±0.00087
500	8.15105±0.00075	10.911±0.010	0.52097±0.00094
1000	9.1558±0.0011	12.0100±0.0070	0.52345±0.00094
5000	11.5016±0.0016	14.949±0.020	0.52667±0.00069
10000	12.5175±0.0024	16.030±0.023	0.52764±0.00071

Table 3.24: Simulation results for path trees

From the above results we can see that the value for C_n is close to the value of $\log_2 n$; in particular under the arbitrary assumption that

$$C_n = \alpha\log_2 n + \beta$$

for $n \geq 500$, then

$$\alpha = 1.00928\pm0.00043; \qquad \beta = -0.8992\pm0.0043$$

References:

[Baer, 75], [Gonnet, 82], [Robson, 82].

3.4.1.6 Heuristic organization schemes on binary trees

When the keys in a binary tree have different accessing probabilities, a randomly generated tree or a balanced tree may not be fully satisfactory. The following heuristic organization schemes offer ways to build better trees when the accessing probabilities are known.

For all these heuristics we will denote by p_i the accessing probability of the ith key. We will denote by q_i the probability of an unsuccessful access, searching for a key with value in between the ith and $i+1^{st}$ keys. In all cases, $\sum_i p_i + \sum_i q_i = 1$. The entropy, or uncertainty of the set of $p_i s$ (or $p_i s$

and $q_i s$), is

$$H(\vec{p}) = -\sum_i p_i \log_2 p_i$$

$$H(\vec{p},\vec{q}) = -\sum_i p_i \log_2 p_i - \sum_i q_i \log_2 q_i$$

Heuristics for known probabilities

The first four algorithms allow a top-down construction, and share the common pseudo-code construction:

Top-down binary tree construction

```
BuildTree( SetOfKeys ) : tree;
    K := select( SetOfKeys );
    A1 := Keys in SetOfKeys < K;
    A2 := Keys in SetOfKeys > K;
    return( NewNode( K, BuildTree(A1), BuildTree(A2) )
    end;
```

(a) *Insert in decreasing probability order.* In this way, the keys most likely to be sought are closer to the root and have shorter search paths. This method requires either a reordering of the keys before they are put into the tree or the selection of the maximum probability at each step. For this analysis, we will assume that the keys are numbered in decreasing probability order, i.e. $(p_1 \geq p_2 \geq ... \geq p_n)$. Then for a random tree

$$C_n = \sum_{i=1}^{n} p_i H_i - 1$$

where $H_i = \sum_{j=1}^{i} 1/j$ is the ith harmonic number.

(b) *Median split.* In this scheme we choose the root so that the total accessing probabilities of both the left and right subtrees are as close as possible to $1/2$. This is repeated recursively on both subtrees. This arrangement gives the information theoretic optimum. For this heuristic

$$C_n^{Opt} \leq C_n^{MS} \leq 2 + 1.44042...H(\vec{p},\vec{q})$$

(c) It is possible to mix approaches (a) and (b). We allow a tolerance δ, and examine the elements for which the accessing probabilities of the left and right subtrees fall into the range $1/2 \pm \delta$. From these elements, we choose the one with the highest accessing probability to be the root. This selection procedure is repeated recursively for the nodes in each subtree. Experimental results indicate that these trees are within 2% to 3% from optimal.

(d) Another way of combining approaches (a) and (b) produces trees which are also called *Median split trees.* At every node we store two keys;

the first one, the "owner" of the node, is the one with highest probability in the subtree, and the second one is the median of all the values in the subtree. The searching algorithm is almost identical to the normal algorithm:

Median split trees: search

```
procedure search( key : typekey; t : tree );
begin
if t=nil then {*** Not Found ***}
              notfound( key )
else if t↑.OwnerKey = key then
              {*** Found ***}
              found( t↑ )
else if t↑.SplitKey < key then     search( key, t↑.right )
              else            search( key, t↑.left )
end;
```

Using this approach we benefit from the advantages of both (a) and (b) above, at the cost of one extra key per node. The "median split" may be interpreted as the statistical median (a key which splits the tree into two subtrees in such a way that both halves are the closest possible to equiprobable) or as the counting median (a key which splits the tree in equal size halves).

(e) *Greedy trees.* This is a heuristic which constructs trees bottom-up. The construction resembles the Huffman encoding algorithm. At each step we select the three consecutive external/ internal/ external nodes which add to the lowest accessing probability. A node is constructed with the two external nodes as direct descendants and the triplet is replaced by a single external node with the sum of the accessing probabilities. Under this heuristic

$$C_n^{GT} \le 2 + 1.81335...H(\vec{p},\vec{q})$$

Self-organizing heuristics

When we do not know the accessing probabilities we may try heuristic organization schemes similar to the Transpose and Move to Front techniques in sequential searching.

(f) *Exchange with parent, (or simple exchange).* The Transpose method can be adapted for trees by exchanging a node with its parent each time the node is accessed. This is achieved by performing a single rotation on the parent node (a left rotation if the searched node is at its right, a right rotation otherwise). This is not a good heuristic, however, as it tends to be very unstable in some cases. For example, if the probability of accessing any key is uniform, $p_i = 1/n$, then this Exchange with parent technique produces a random binary tree and

$$C_n^{EP} \approx \sqrt{\pi n} - 3 + O(n^{-1/2})$$

(g) *Move to Root.* Corresponding to the Move to Front scheme in linear searching, we have the technique of moving an accessed element to the root. This is achieved, while maintaining the lexicographical order of the tree, by several single rotations on the ancestors of the accessed element. With this Move to Root approach we have

$$C_n^{MR} = 1+2 \sum_{1 \le i < j \le n} \frac{p_i p_j}{p_i + ... + p_j} \le 2\ln(2)H(\vec{p})+1$$

$$\sigma^2(A_n^{MR}) \le 2\ln n + O(1)$$

(h) *Dynamic trees, (or D-trees).* Dynamic trees use a self-organizing technique based on estimating the accessing probabilities by keeping counters for the number of successful/unsuccessful searches at each internal/external node. The tree is balanced with respect to these counters, like the balance done for $BB[\alpha]$ trees (3.4.1.4). If f_i denotes the relative accessing frequency of node i, then the number of access needed to locate node i is $O(\log(1/f_i))$.

Shape heuristics

(i) *Fringe reorganization.* This type of heuristics guarantees that any subtree with size k or smaller is of minimal height (or, equivalently, of minimal internal path). The simplest heuristic is for $k=3$ which reorganizes any subtree with 3 nodes which is not in perfect balance. Under random insertions, a tree constructed using $k=3$ will have

$$C_n' = \frac{12}{7}H_{n+1} - \frac{75}{49} \approx 1.18825...\log_2 n - 0.54109... \qquad \text{for } n \ge 6$$

$$\sigma^2(A_n') = \frac{300}{343}H_{n+1} - \frac{144}{49}H_{n+1}^{(2)} + \frac{5056}{2401} + \frac{2304}{343(n+1)n(n-1)\cdots(n-5)}$$

for $n \ge 13$.

In general, if $k=2^t-1$ then

$$C_n' = \frac{H_n}{H_{2t}-H_t} + O(1)$$

$$\sigma^2(A_n') = \frac{H_{2t}^{(2)} - H_t^{(2)}}{(H_{2t}-H_t)^3}H_n + O(1)$$

References:

[Allen *et al.*, 76], [Allen, 82], [Baer *et al.*, 77], [Comer, 80], [Fredman, 75], [Gotlieb *et al.*, 72], [Horibe *et al.*, 79], [Itai *et al.*, 80], [Knuth, 73], [Korsh, 81], [Korsh, 82], [Martin *et al.*, 72], [Mehlhorn, 75], [Mehlhorn, 77], [Mehlhorn, 79], [Poblete *et al.*, 82], [Sheil, 78], [Sleator *et al.*, 83], [Walker *et al.*, 76].

3.4.1.7 Optimal binary tree search

When we want to minimize the average case search and all the nodes in the tree are equally probable, or when we want to minimize the worst case, it is easy to see that the optimal tree is the one with minimum height. Equivalently, such an optimal tree has all its leaves at, at most, two consecutive levels.

When the nodes in the tree have different accessing probabilities, and these probabilities are known, we can construct an optimal (minave) search tree. For these optimal trees,

$$H(\vec{p}) - \log_2(eH(\vec{p})) + 1 \leq C_n^{Opt} \leq H(\vec{p}) + 1$$

$$H(\vec{q}) \leq C_n^{Opt} \leq 2 + H(\vec{q})$$

if $p_i = 0$.

The following algorithm constructs an optimal tree given the probabilities of successful searches (p_i) and the probabilities of unsuccessful searches (q_i). This algorithm due to Knuth uses a dynamic programming approach, computing the cost and root of every tree composed of contiguous keys. To store this information, the algorithm uses two upper-triangular matrices dimensioned $n \times n$. Both its storage and time requirements are $O(n^2)$.

Optimal binary tree construction (Knuth)

```
function OptTree( keys : ArrayKeys; p : ArrayCost; q : ArrayCost ) : tree;

var     wk, wki, min : cost;
        i, ik, indxmin, j, k : integer;
        {*** r[i,j] indicates the root of the optimal tree formed
                with keys from i to j ***}
        r : array[0..n,0..n] of integer;
        {*** c[i,j] indicates the optimal cost of the tree with
                keys from i to j ***}
        c : array[0..n,0..n] of cost;

        function CreateTree( i, j : integer ) : tree;
        {*** Create optimal tree from information in r[i,j] ***}
        var     t : tree;
        begin
        if i=j then     CreateTree := nil
        else    begin
                new(t);
                t↑.k := keys[ r[i,j] ];
                t↑.left := CreateTree( i, r[i,j]-1 );
                t↑.right := CreateTree( r[i,j], j );
                CreateTree := t
                end
        end;
```

```
begin
{*** Initializations ***}
c[0,0] := q[0];
for i:=1 to n do begin
        c[i,i] := q[i];
        c[i-1,i] := 2*( q[i-1] + q[i] ) + p[i];
        r[i-1,i] := i
        end;

{*** Main loop to compute r[i,j] ***}
wk := q[0];
for k:=2 to n do begin
        wk := wk + q[k-1] + p[k-1];
        wki := wk;
        for i:=0 to n-k do begin
                ik := i+k;
                wki := wki + q[ik] + p[ik];
                min := maxint;
                {*** Select root with lowest cost ***}
                for j:=r[i,ik-1] to r[i+1,ik] do
                        if c[i,j-1]+c[j,ik] < min then begin
                                min := c[i,j-1]+c[j,ik];
                                indxmin := j
                                end;
                c[i,ik] := min + wki;
                r[i,ik] := indxmin;
                wki := wki - q[i] - p[i+1];
                end
        end;

OptTree := CreateTree( 0, n );
end;
```

If we are interested in the unsuccessful probabilities alone ($p_i = 0$), the *Hu-Tucker* algorithm will construct an optimal tree in $O(n \log n)$ time and $O(n)$ space.

References:
[Allen, 82], [Bayer, 75], [Choy *et al.*, 77], [Choy *et al.*, 78], [Garey, 74], [Garsia *et al.*, 77], [Hosken, 75], [Hu *et al.*, 71], [Hu *et al.*, 72], [Hu *et al.*, 79], [Hu, 73], [Itai, 76], [Kennedy, 72], [Kleitman *et al.*, 81], [Knuth, 71], [Knuth, 73], [Mehlhorn, 81], [Wessner, 76], [Wikstrom, 79].

3.4.1.8 Rotations in binary trees

Rotations in binary trees are operations that modify the structure (shape) of the tree without altering the lexicographical ordering in the tree. These transformations are very useful in keeping the tree structure balanced.

The simplest rotation, which is usually called *single rotation,* is illustrated by the next figure.

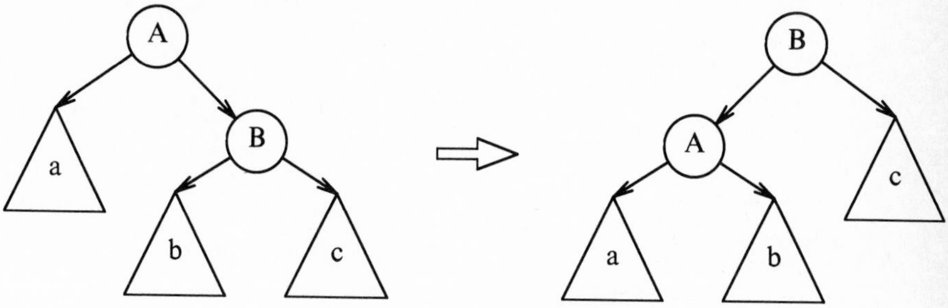

Figure 3.25: Single left rotation

There are two possible such situations, the one shown in figure 3.25 and its symmetric which are called *left* and *right* single rotations respectively. The procedures to perform these rotations are

Single left rotation

```
procedure lrot( var t : tree );
var     temp : tree;
begin
        temp := t;
        t := t↑.right;
        temp↑.right := t↑.left;
        t↑.left := temp;
end;
```

Single right rotation

```
procedure rrot( var t : tree );
var     temp : tree;
begin
        temp := t;
        t := t↑.left;
        temp↑.left := t↑.right;
        t↑.right := temp;
end;
```

A *double rotation* is a more complicated transformation illustrated in figure 3.26.

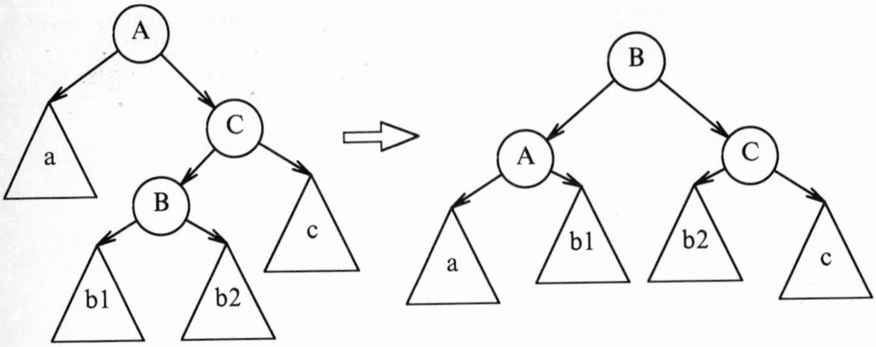

Figure 3.26: Double left rotation

Figure 3.26 illustrates a transformation called double *left* rotation. Its symmetric is called a double *right* rotation. Both rotations can be described in terms of two single rotations, for example a double left rotation at the node pointed by t is achieved by

Double left rotation

---

rrot(t↑.right); *lrot(t);*

In many cases the nodes carry some information about the balance of their subtrees. For example, in the AVL trees, each node contains the difference in height of its subtrees; in the weight balanced trees each node contains the total number of nodes in its subtree. This information should be reconstructed by the single rotation, and consequently double rotations or more complicated rotations based on single rotations do not need to reconstruct any information.

Let *bal* contain the difference in height between the right subtree and the left subtree (h(t↑.right) - h(t↑.left)), as in AVL trees.

For example, after a single left rotation, the new balance of the nodes A and B (Fig. I) is given by:

$$NewBal(A) = OldBal(A) - 1 - \max(OldBal(B), 0)$$

$$NewBal(B) = \min(OldBal(A)-2, OldBal(A)+OldBal(B)-2, OldBal(B)-1)$$

The complete code for a single left rotation becomes

Single left rotation

```
procedure lrot( var t : tree );
var     temp : tree;
        a : integer;
begin
        temp := t;
        t := t↑.right;
        temp↑.right := t↑.left;
        t↑.left := temp;
        {*** adjust balance ***}
        a := temp↑.bal;
        temp↑.bal := a - 1 - max( t↑.bal, 0 );
        t↑.bal := min( a-2, a+t↑.bal-2, t↑.bal-1);
end;
```

3.4.1.9 Deletions in binary trees

The operation of deleting a node in a binary tree is relatively simple if the node to be deleted has a null descendant. In this case the node is replaced by the other descendant. If both descendants are non-null the node has to be moved down the tree until it has a non-null descendant.

One way of moving the node to the fringe of the tree is to swap it with one of its lexicographically ordered neighbours. Another strategy, better suited for balanced trees, is to gradually move the node towards the fringe by the use of rotations.

The following procedure performs deletions on weight balanced or path balanced trees.

Deletions on weight balanced trees

```
procedure delete( key : typekey; var t : tree );

begin
if t = nil then        Error {*** key not found ***}

{*** search for key to be deleted ***}
else if t↑.k < key then      delete( key, t↑.right )
else if t↑.k > key then delete( key, t↑.left )
```

```
{*** key found, delete if a descendant is nil ***}
else if t↑.left  = nil then t := t↑.right
else if t↑.right = nil then t := t↑.left

{*** no descendant is null, rotate on heavier side ***}
else if wt( t↑.left ) > wt( t↑.right ) then begin
        {*** left side is heavier, do a right rotation ***}
        if wt( t↑.left↑.left ) < wt( t↑.left↑.right ) then
                lrot( t↑.left );
        rrot( t );
        delete( key, t↑.right )
        end
else    {*** do a left rotation ***}
        begin
        if wt( t↑.right↑.left ) > wt( t↑.right↑.right ) then
                rrot( t↑.right );
        lrot( t );
        delete( key, t↑.left )
        end;

{*** reconstruct weight information ***}
if t≠nil then t↑.weight := wt( t↑.left ) + wt( t↑.right );
end;
```

For height balanced (AVL) trees we simply replace the function *wt( )* by the height of the subtree.

3.4.2 B-trees

A B-tree is a balanced multiway tree with the following properties:

(a) every node has at most $2m+1$ descendants;

(b) every internal node except the root has at least $m+1$ descendants, the root either being a leaf or having at least two descendants;

(c) the leaves are null nodes which all appear at the same depth.

B-trees are usually named after their allowable branching factors, i.e. $m+1$-$2m+1$. For example, 2-3 trees are B-trees with $m=1$; 6-11 trees are B-trees with $m=5$. B-trees are used mainly as a primary key access method for large databases which cannot be stored in internal memory. Recall the definition of multiway trees,

$$\textbf{mt}-\textbf{N}-\textbf{D}-\textbf{LEAF} : [\, \textbf{int}, \{\textbf{D}\}_1^{\textbf{N}}, \{\textbf{mt}-\textbf{N}-\textbf{D}-\textbf{LEAF}\}_0^{\textbf{N}}] ; \textbf{LEAF}.$$

then the data structure for a general B-tree is $mt-2m-D-nil$. For our C algorithms we will use the definition:

B-tree data structure

```
typedef struct btnode {              /*** B-Tree Definition ***/
        int d;                       /*** number of active entries ***/
        typekey k[2*M];              /*** Keys ***/
        struct btnode *p[2*M+1];     /*** Pointers to subtrees ***/
        } node, *btree;
```

Note that, in C, arrays always start with index 0, consequently the array containing the keys runs from 0 to 2*M−1. The lexicographical order is given by the fact that all the keys in the subtree pointed by $p[i]$ are greater than $k[i-1]$ and less than $k[i]$.

Let E_n and E_n' represent the number of nodes accessed in successful and unsuccessful searches respectively. Let $h(n)$ be the height of a B-tree with n keys. Then

$$1 \le E_n \le h(n)$$

$$E_n' = h(n)$$

$$E[E_n] = h(n) - \frac{1}{2m\ln 2} + O(m^{-2})$$

$$\lceil \log_{2m+1}(n+1) \rceil \le h \le 1 + \lfloor \log_{m+1}((n+1)/2) \rfloor$$

$$\sigma^2(E_n) \le \frac{m+1}{m^2}$$

Let t_n be the number of different B-trees with n leaves. We have

$$B(z) = \sum_{n=0}^{\infty} t_n z^n = B(P(z)) + z$$

where $P(z) = \dfrac{z^{m+1}(z^{m+1}-1)}{z-1}$,

$$t_n = \frac{\phi_m^{-n}}{n} Q(\log n)(1 + O(n^{-1}))$$

where $0 < \phi_m < 1$ and ϕ_m is a root of $P(z) = z$ and $Q(x)$ is a periodic function in x with average value $\dfrac{\phi_m}{\ln P'(\phi_m)}$ and period $\ln P'(\phi_m)$.

m	ϕ_m	$\ln P'(\phi_m)$	$\phi_m / \ln P'(\phi_m)$
1	0.61803...	0.86792...	0.71208...
2	0.68232...	1.01572...	0.67176...
5	0.77808...	1.21563...	0.64006...
10	0.84439...	1.34229...	0.62907...

Table 3.27: Parameters for counting different B-trees

$$\phi_m = 1 - \frac{w(m)}{m} + \frac{w(m)+2}{2w(m)+2}\left(\frac{w(m)}{m}\right)^2 + O((w(m)/m)^3)$$

where $w(m)e^{w(m)}=m$, and

$$t_n = O\left(\frac{e^{w(m)n/m}}{n}\right)$$

Let N_n be the expected number of nodes in a randomly generated B-tree with n keys. Then

$$\frac{2m+1}{4m(m+1)(H_{2m+2}-H_{m+1})} \le \frac{N_n}{n} \le \frac{1}{2m(H_{2m+2}-H_{m+1})}$$

$$\frac{N_n}{n} = \frac{1}{2m\ln 2} + O(m^{-2})$$

Let S_n be the average number of node-splits produced by an insertion into a randomly generated B-tree with n keys.

$$\frac{1}{2(m+1)(H_{2m+2}-H_{m+1})} \le S_n$$

Below we present a description of the algorithm for searching B-trees. Note that in this case we can convert the "tail recursion" into an iteration very easily.

B-tree search

```
search( key, t )
typekey key;
btree t;

{       int i;
while (t ≠ NULL) {
        for (i=0; i<t→d && key>t→k[i]; i++);
        if (key == t→k[i])
                { found( t, i );        return; }
        t = t→p[i];
        }
notfound( key );
};
```

B-tree insertion

```
btree insert( key, t )
typekey key;
btree t;

{
typekey ins;
extern btree NewNode;
      ins = InternalInsert( key, t );
      /*** check for growth at the root ***/
      if ( ins ≠ NoKey )  return( CreateNode( ins, t, NewNode) );
      return( t );
};

  typekey InternalInsert( key, t )
  typekey key;
  btree t;
  extern btree NewNode;
  {int i, j;
  typekey ins;
  btree tempr;
  if ( t == NULL ) {      /*** The bottom of the tree has been reached:
                          indicate insertion to be done ***/
            NewNode = NULL;
            return( key );
            }
  else {
       for ( i=0; i<t→d && key>t→k[i]; i++ );
       if ( key == t→k[i] )
            Error; /*** Key already in table ***/
       else  {
            ins = InternalInsert( key, t→p[i] );
            if ( ins ≠ NoKey )
            /*** the key in "ins" has to be inserted in present node ***/
                  if (t→d < 2*M)   InsInNode( t, ins, NewNode );
                  else   /*** Present node has to be split ***/
                         {/*** Create new node ***/
                         if ( i≤M )   {
                                tempr = CreateNode( t→k[2*M-1], NULL, t→p[2*M] );
                                t→d--;
                                InsInNode( t, ins, NewNode );
                                }
                         else tempr = CreateNode( ins, NULL, NewNode );
                         /*** move keys and pointers ***/
                         for ( j=M+2; j≤2*M; j++ )
                                InsInNode( tempr, t→k[j-1], t→p[j] );
                         t→d = M;
                         tempr→p[0] = t→p[M+1];
                         NewNode = tempr;
                         return( t→k[M] );
                         }
            }
       return( NoKey );
       }
  };
```

The above algorithm is structured as a main function *insert* and a subordinate function *InternalInsert*. The main function handles the growth at the root, while the internal handles the recursive insertion in the tree.

The insertion function returns a pointer to the resulting tree. This pointer may point to a new node when the B-tree grows at the root.

The insertion algorithm uses the global variable *NewNode* to keep track of newly allocated nodes in the case of node-splitting. The function *InsInNode* inserts a key and its associated pointer in lexicographical order in a given node. The function *CreateNode* allocates storage for a new node and inserts one key and its left and right descendant pointers. The value *NoKey* is an impossible value for a key and it is used to signal that there is no propagation of splittings during an insertion.

Although B-trees can be used for internal memory dictionaries, this structure is most suitable for external searching. For external dictionaries, each node can be made large enough to fit exactly into a physical record, thus yielding, in general, high branching factors. This produces trees with very small height.

B-trees are well suited to searches which look for a range of keys rather than one unique key. Furthermore, since the B-tree structure is kept balanced during insertions and deletions, there is no need for periodic reorganizations.

Several variations have been proposed for general B-trees with the intention of improving the utilization factor of the internal nodes. Note that a better storage utilization will result in a higher effective branching factor, shorter height and less complexity. The variations can be loosely grouped in three different classes:

Overflow techniques (or B-trees)*
These are variations which try to prevent the splitting of nodes. When an overflow occurs during an insertion, instead of splitting the node we can:

(a) scan a right or left brother of the node to see if there is any room, and, if there is, we can transfer one key-pointer pair (the leftmost or rightmost respectively) to make room in the overflowed node;
(b) scan both left and right siblings of a node;
(c) scan all the descendants of the parent of the node.

If a splitting is still necessary, the new nodes may take some keys from their siblings to achieve a more even distribution of keys in nodes.

Variable-length array implementations
These variations replace the arrays used to store keys and pointers at every node for some other structure which allows variable length, and may save space when the node is not full. For example we could use a linked list where each node in the list contains a key and a pointer to the subtree at its left and the last pointer of the list points to the rightmost subtree. The sequence, in this case, is defined by:

$$\mathbf{s-D : [\ KEY, [D], s-D\ ] ; [D]}$$

Each node in the B-tree contains one of these sequences. These sequences can be viewed as restricted binary trees, with two types of pointers: vertical pointers (those which point to nodes down the tree) and horizontal pointers (those pointing at the next link of the list). This last view is the topic of section 3.4.2.2.

These variations have to be able to store variable length records to be successful.

When the keys are themselves of variable length, we can slightly relax the conditions on B-trees and require that each node be between 50% and 100% full, without any explicit reference to the actual number of keys stored.

Let m be the total number of characters that can be stored in a node, and let k be the maximum size of a key. Then we can guarantee that the number of characters per node will be between $\lfloor (m+1)/2 \rfloor - k$ and m.

Index B-trees
(Sometimes called *B+-trees* or *B*-trees*). The idea behind these trees is to move all the data which is normally associated with a record to the leaves of the tree. The internal nodes contain only keys which are used to direct the searching; the complete records reside at the leaves. The keys in the internal nodes may not even belong to the file. Typically the leaves are pointers to external buckets of uniform size b. The data structure is now represented as:

$$\mathbf{mt-N-D-LEAF \rightarrow mt-2m-KEY-[D_1^b]}.$$

The above variations are somewhat orthogonal, in the sense that these can be applied simultaneously to achieve varying degrees of optimization. Note that the limits of the range for any gain in efficiency are from about 70% occupation (for randomly generated trees) to 100% occupation (optimal trees). The coding complexity of some of these implementations may not justify the gains.

Table 3.28 presents simulation results of 6-11 trees for several sizes, and the next table shows simulation results for various branching factors and a constant size. In both cases, E_n indicates the number of nodes accessed, $h(n)$ indicates the height of the tree, N_n is the average number of nodes per key in the tree, and S_n is the average number of splits that the $n+1^{st}$ insertion will require.

n	$E[E_n]$	$E[h(n)]$	N_n	S_n
5	1	1	0.2	0
10	1	1	0.1	1
50	1.88963±0.00012	2.00000±0.00007	0.1537±0.00012	0.1272±0.00150
100	2.8319±0.0026	2.9603±0.0027	0.15799±0.00015	0.1402±0.0021
500	2.86006±0.00013	3.00000±0.00075	0.14593±0.00013	0.1354±0.0017
1000	3.85715±0.00013	4.0000±0.0015	0.14685±0.00013	0.1392±0.0018
5000	3.870±0.012	4.015±0.012	0.14580±0.00014	0.1474±0.0020
10000	4.85450±0.00013	5.000±0.015	0.14600±0.00013	0.1467±0.0019

Table 3.28: Simulation results for 6-11 trees

type	$E[E_n]$	$E[h(n)]$	N_n	S_n
2-3	10.25486±0.00041	11.000±0.015	0.74622±0.00041	0.7444±0.0029
6-11	4.85450±0.00013	5.000±0.015	0.14600±0.00013	0.1467±0.0019
11-21	3.92754±0.00010	4.000±0.015	0.07286±0.00010	0.0755±0.0021
21-41	2.96388±0.00007	3.000±0.015	0.03642±0.00007	0.0381±0.0018
51-101	2.98603±0.00005	3.000±0.015	0.01427±0.00005	0.0136±0.0016

Table 3.29: Simulation results for B-trees with 10000 keys

The simulation results indicate that the variance on the number of nodes accessed is very small. Induced by the formula for the upper bound on the variance, and with the arbitrary assumption that

$$\sigma^2(E_n) = \alpha \frac{m+1}{m^2} + \beta$$

for $n = 10000$ we find that

$$\alpha = 0.6414 \pm 0.0005; \qquad \beta = 0.0053 \pm 0.0005$$

References:
[Aho *et al.*, 83], [Batory, 81], [Bayer *et al.*, 72], [Bayer *et al.*, 77], [Bayer, 71], [Comer, 79], [Culik *et al.*, 81], [Gotlieb *et al.*, 78], [Gotlieb, 81], [Guibas *et al.*, 78], [Guibas, 77], [Gupta *et al.*, 83], [Hansen, 81], [Held *et al.*, 78], [Horowitz *et al.*, 76], [Huddleston *et al.*, 81], [Huddleston *et al.*, 82], [Knuth, 73], [Lomet, 78], [Lomet, 81], [Maly, 78], [McCreight, 77], [Odlyzko, to app.], [Ottmann *et al.*, 81], [Ottmann *et al.*, 82], [Ouksel *et al.*, 81], [Robinson, 81], [Rosenberg *et al.*, 78], [Rosenberg *et al.*, 81], [Samadi, 76], [Snyder, 78], [Standish, 80], [Strong *et al.*, 79], [Tamminen, 83], [van-Leeuwen *et al.*, 83], [Wirth, 76].

3.4.2.1 2-3 trees

2-3 trees are the special case of B-trees when $m = 1$. Each node has two or three descendants, and all the leaves are at the same depth.

$$\lceil \log_3 n+1 \rceil \le h(n) \le \lfloor \log_2 n+1 \rfloor$$

Let t_n be the number of different 2-3 trees with n leaves. Then

$$B(z) = \sum_{n=0}^{\infty} t_n z^n = B(z^2+z^3)+z$$

$$t_n = \frac{\phi^n}{n} Q(\ln n)(1+O(n^{-1}))$$

where $\phi=(1+\sqrt{5})/2$ is the "golden ratio", and $Q(x)$ is a periodic function with period $\ln(4-\phi)$ and mean value $(\phi \ln(4-\phi))^{-1}$.

Let N_n be the expected number of nodes in a 2-3 tree built by the insertion of a random permutation of n keys. Then

$$0.72683...+O(n^{-4.37}) \le \frac{N_n}{n} \le 0.76555...+O(n^{-4.37})$$

Let S_n be the number of node-splits produced by an insertion into a random 2-3 tree with n keys, then

$$0.68810...+O(n^{-4.37}) \le S_n \le 0.45574...+ 0.07745... \times \lfloor \log_2(n+1) \rfloor + O(n^{-4.37})$$

If S_n converges when $n \rightarrow \infty$ then

$$S_\infty \le 0.76555...$$

If we assume all trees of height h are equally likely, then

$$N_n = (0.48061...)3^h$$

$$E[keys] = (0.72161...)3^h$$

The algorithm for searching and performing insertions in 2-3 trees is the same as the general algorithm for B-trees with $m=1$.

As opposed to general B-trees, 2-3 trees are intended for use in main memory.

Below in table 3.30, we give figures showing the performance of 2-3 trees constructed from random sets of keys.

n	$E[E_n]$	$E[h(n)]$	N_n	S_n
5	1.68	2	0.72	0.40
10	2.52857±0.00020	3	0.77143±0.00020	0.52218±0.00096
50	4.1887±0.0033	4.8479±0.0035	0.75617±0.00045	0.7169±0.0030
100	4.7158±0.0065	5.4088±0.0068	0.74701±0.00049	0.7503±0.0032
500	6.464±0.012	7.196±0.012	0.74594±0.00045	0.7466±0.0032
1000	7.2788±0.0055	8.0165±0.0056	0.74570±0.00045	0.7477±0.0033
5000	9.2588±0.0049	10.0025±0.0049	0.74568±0.00045	0.7485±0.0033
10000	10.25486±0.00041	11.000±0.015	0.74622±0.00041	0.7444±0.0029

Table 3.30: Simulation results for 2-3 trees

2-3 Brother trees

2-3 Brother trees are 2-3 trees with the additional constraint that a binary node has to have ternary brothers. With this restriction it is still possible, albeit complicated, to update a tree in $O(\log n)$ time. Let N_n^B be the number of nodes and $h^B(n)$ the height of a 2-3 Brother tree with n keys. Then

$$\lceil \log_3(n+1) \rceil \leq h^B(n) \leq \lfloor 0.78644...\log_2 n - 0.39321... \rfloor$$

$$\frac{1}{2} \leq \frac{N_n^B}{n} \leq \frac{1}{\sqrt{2}} = 0.70710...$$

$$1 \leq \frac{E[N_n^B]}{n} \leq 1.4142...$$

References:

[Aho *et al.*, 74], [Bent *et al.*, 80], [Bent, 82], [Bitner *et al.*, 81], [Brown *et al.*, 78], [Brown *et al.*, 80], [Brown, 79], [Brown, 79], [Gupta *et al.*, 82], [Kosaraju, 81], [Kriegel *et al.*, 78], [Kriegel *et al.*, 83], [Larson *et al.*, 79], [Maier *et al.*, 81], [Mehlhorn, 82], [Miller *et al.*, 79], [Olivie, 80], [Reingold, 79], [Rosenberg *et al.*, 78], [Vaishnavi *et al.*, 79], [Yao, 78], [Zaki *et al.*, 79].

3.4.2.2 Symmetric binary B-trees

Symmetric binary B-trees (or SBB trees) are implementations of 2-3 trees. A 2-3 tree node with a single key is mapped into a binary tree node directly; a 2-3 node with two keys is mapped into two nodes as indicated in figure 3.31.

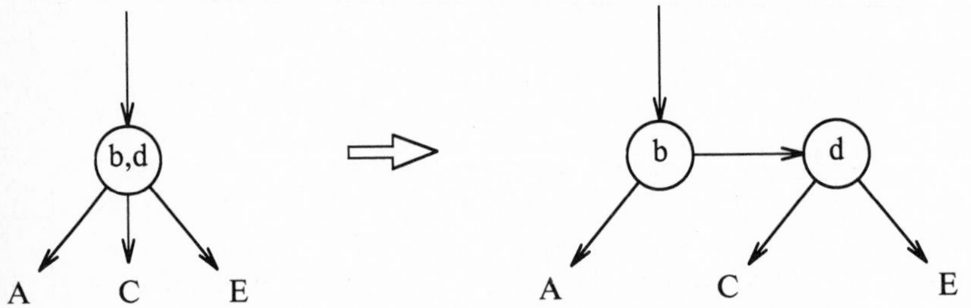

Figure 3.31: Transformation of 2-3 node into an SBB node

SBB trees are binary search trees in which the right and left pointers may be either *vertical* (normal) pointers or *horizontal* pointers. In an SBB tree all paths have the same number of vertical pointers (as in a true B-tree). All nodes except the leaves have two descendants and there are no two consecutive horizontal pointers in any path. In order to maintain the SBB tree property one bit per node is needed to indicate whether the incoming pointer is horizontal or vertical.

Random retrievals, insertions and deletions of keys in an SBB tree with n keys can be done in time of $O(\log n)$. If we let $k(n)$ be the maximum number of keys in any path and $h(n)$ be the height of the SBB tree (calculated by counting only vertical pointers plus one), we have

$$h(n) \leq k(n) \leq 2h(n)$$

$$\lceil \log_2(n+1) \rceil \leq k(n) \leq 2\lfloor \log_2(n+2) \rfloor - 2$$

Let S_n indicate the number of split transformations (a split transformation for SBB trees is similar to a rotation for AVL trees) required during the insertion of the nth key into a random tree. Let HI_n indicate the number of local height increase transformations required during the insertion of the nth key into the tree. Then

$$0.35921... + O(n^{-5}) \leq E[S_n] \leq 0.55672... + O(n^{-5})$$

$$E[HI_n] = \frac{23}{35} + O(n^{-5})$$

Let V_n be the number of nodes visited to process n random insertions/deletions into the empty tree. Then

$$E[V_n] \leq \frac{5}{2}n + O(n^{-5})$$

Table 3.32 shows some simulation results for SBB trees. C_n is the average number of nodes visited during a successful search and S_n, V_n and $h(n)$ have the meaning described above.

n	C_n	S_n	V_n	$E[h(n)]$
5	2.2000±0.0003	0.213±0.023	1.213±0.023	3.000±0.020
10	2.9057±0.0035	0.293±0.015	1.663±0.021	4.023±0.022
50	4.9720±0.0051	0.3594±0.0050	2.1692±0.0073	7.009±0.016
100	5.9307±0.0054	0.3733±0.0046	2.2757±0.0072	8.093±0.033
500	8.2419±0.0059	0.3868±0.0027	2.3801±0.0047	11.027±0.026
1000	9.2537±0.0062	0.3872±0.0023	2.3975±0.0042	12.140±0.068
5000	11.6081±0.0073	0.3876±0.0013	2.4088±0.0023	15.014±0.028
10000	12.6287±0.0083	0.3880±0.0011	2.4109±0.0019	16.180±0.108

Table 3.32: Simulation results for SBB trees

From the simulation results we can see that the value for C_n is close to the value of $\log_2 n$, in particular, under the arbitrary assumption that

$$C_n = \alpha \log_2 n + \beta$$

then

$$\alpha = 1.0186 \pm 0.0010; \qquad \beta = -0.909 \pm 0.011$$

While every AVL tree can be transformed into an SBB tree, the converse is not true. Thus the class of AVL trees is a proper subclass of the SBB trees. Experimental results show that, on the average, SBB trees perform approximately as well as AVL trees. Indeed SBB trees require less work than AVL trees to maintain balance, but this is at the expense of search time. (The search time is only slightly longer and the maintenance time is in some areas significantly less.) As a practical structure SBB trees should be considered as an option for representing dictionaries.

References:
[Bayer, 72], [Olivie, 80], [Olivie, 80], [Tarjan, 83], [Ziviani *et al.*, 82].

3.4.2.3 1-2 trees

1-2 trees are a special case of B-trees in which every node can have either one key or no keys. Consequently, every node has either 2 or 1 descendants. A node with no keys and one descendant is called a *unary* node. Since we allow nodes without keys, some additional restrictions are usually imposed so that a tree containing n keys is of bounded size (number of nodes).

1-2 Brother trees

1-2 Brother trees are 1-2 trees with the additional constraint that every unary node has a binary brother.

There is a close correspondence between 1-2 Brother trees and AVL trees, as any 1-2 Brother tree can be easily converted into an AVL tree and vice-versa. This correspondence is a very natural one and consists in deleting the unary nodes (brother→avl) or inserting a unary node on the shorter sub-tree (avl→brother) of every node. Moreover, for some methods of insertion and deletion, any sequence of operations (insertions/deletions) on AVL trees and 1-2 Brother trees will produce equivalent trees.

Let N_n^B be the number of nodes in a 1-2 Brother tree with n keys, then for a tree constructed from a set of random keys:

$$n \le N_n^B \le 1.61803...n$$

$$\frac{40n-5}{35} \le E[N_n^B] \le \frac{156n-19}{105}$$

for $n \ge 6$.

1-2 Son trees

1-2 Son trees are 1-2 trees with the additional constraint that in no parent-descendant pair are both nodes unary. There is a close correspondence between 1-2 Son trees and SBB trees as any Son tree can be converted to an SBB tree and vice-versa. With this restriction, letting N_n^S denote the number of nodes used by a tree with n keys

$$n \le N_n^S \le 3n+1$$

$$\frac{48n+13}{35} \le E[N_n^S] \le \frac{72n-33}{35}$$

for $n \ge 6$.

1-2 Neighbour trees

Neighbour trees of order k are 1-2 trees with the additional constraint that every unary node has at least one right neighbour and its first k right neighbours, if these exist, are binary. For these trees the height is bounded by

$$\lfloor \log_2(n+1) \rfloor \leq h(n) \leq \frac{\log_2 n}{\log_2(2-1/(k+1))} + 1$$

References:
[Maurer *et al.*, 76], [Mehlhorn, 82], [Olivie, 80], [Olivie, 81], [Ottmann *et al.*, 78], [Ottmann *et al.*, 79], [Ottmann *et al.*, 80], [Ottmann *et al.*, 80], [Ottmann *et al.*, 80], [Ottmann *et al.*, 81], [Ottmann *et al.*, 82].

3.4.3 Index and indexed sequential files

An *indexed file* is a superimposition of a dictionary structure called the *main file* upon another dictionary structure called the *index file.* The index file is constructed on a subset of the keys of the main file. Using our notation for data structures, a single level index is defined by:

main−file(KEY) : SET(bucket(KEY))

index(KEY) : DICT(KEY,[bucket(KEY)])

bucket(KEY) : DICT(KEY);

In the above definition, **DICT** stands for the organization of the index file, while the collection of all the **bucket(KEY)** is the main file.

Indexed files can be organized in several levels. By adding an index of the index we increase the number of levels by one. This is formally described by mapping the **bucket(KEY)** to

bucket(KEY) : index(KEY)

instead. If the same **DICT** structures for each level of indexing are chosen, the file has *homogeneous indexing.* In practice, the number of levels is very small and homogeneous (typically 1 or 2 levels).

The typical choices for the **DICT** structure in the index file are arrays and trees. The typical choice for the **bucket** is a sequential array. An indexed file can, however, be implemented using any selection for the **DICT** structures in the index file and bucket and the **SET** representation for the main file. Normally the following constraints are imposed on the structure:

(a) each index entry contains as key the maximum key appearing in the pointed **bucket(KEY);**

(b) the index file structure should perform range searches, or nearest-neighbour searches efficiently, the type of search of most interest being "search the smallest key $\geq X$";

(c) the **bucket(KEY)** should allow some type of dynamic growth (overflow records, chaining, etc.), which should not be of bounded size;

(d) it should be possible to scan all the components in a bucket sequentially and all the components of the set sequentially, or, in other words, it should be possible to scan all the main file sequentially;

(e) the index contains an artificial key ($-\infty$) which is larger than any other key in the file.

Searching an array index

```
function SearchIndex( key : typekey ) : BucketAddress;
var     low, high, j : integer;

begin
low := 0;
high := n;    {*** highest index entry ***}
while high-low > 1 do begin
        j := (high+low) div 2;
        if key ≤ index[j].k then   high := j
                else               low := j
        end;
SearchIndex := index[high].BuckAddr
end;
```

Searching array buckets with overflow

```
procedure SearchBucket( key : typekey; p : BucketAddress );
label  999;
var    i : integer;

begin
while p ≠ nil do begin
        read-bucket(p) into bucket;
        i := B;
        while (i>1) and (bucket.r[i].k>key) do i := i-1;
        if bucket.r[i].k = key then goto 999  {*** break ***}
        else if i=B then       p := bucket.next
                       else    p := nil
        end;
999:
if p ≠ nil then        found( bucket.r[i] )
        else           notfound( key )
end;
```

The goal of indexed files is to have an index small enough to keep in main memory, and buckets small enough to read with a single access. In this ideal situation, only one external access per random request is needed.

B*-trees (section 3.4.2) are a generalization of a special implementation of index files.

Searching a single-level indexed file

SearchBucket(key, SearchIndex(key));

Typically the index part of the file is considered to be a fixed structure and no updates are performed on it. In case the file grows or shrinks or alters its distribution significantly, it is easier to reconstruct the index entirely.

3.4.3.1 Index sequential access method

A particular implementation of indexed files are the ISAM (Index Sequential Access Method) files. For these files the index file and set are both arrays. The buckets are composed of an array of records of fixed maximum size and an additional pointer to "overflow" buckets. Since the index file and main file are both arrays, there is no need to keep pointers in the index. The array index in the index file corresponds to the array index (bucket index) on the main file.

$$\mathbf{index(KEY)} : \{\mathbf{KEY}\}_1^N$$

$$\mathbf{main-file} : \{\mathbf{bucket(KEY)}\}_1^{N+W};$$

$$\mathbf{bucket(KEY)} : (\{\mathbf{KEY, D}\}_1^B, \mathbf{int});$$

In the above definition, **B** is the bucket size, **N** denotes the number of buckets in the main file, and **W** denotes the number of buckets reserved for overflow. The integer in the **bucket(KEY)** is the index of the corresponding overflow bucket.

The buckets are designed to closely match the physical characteristics of devices, e.g. typically a bucket fully occupies a track in a disk. In some cases the index is organized as an indexed file itself, in which case the ISAM becomes a two-level index. For two-level indices the same array structures are used. The top level index is made to match a physical characteristic of the device, for example, a cylinder in a disk.

General References:
[Batory, 81], [Gairola *et al.*, 76], [Ghosh, 69], [Keehn *et al.*, 74], [Larson, 81], [Leipala, 81], [Leipala, 82], [Lum, 70], [Maruyama *et al.*, 77], [Mullin, 72], [Nakamura *et al.*, 78], [Shneiderman, 74], [Strong *et al.*, 79], [Whitt *et al.*, 75], [Zvegintzov, 80].

3.4.4 Digital trees

Digital trees or *tries* are recursive tree structures which use the characters, or digital decomposition of the key, to direct the branching. The name trie comes from the word re*trie*val. A node in a trie is either an external node and contains one record, or it is an internal node and contains an array of pointers to nodes or null pointers. The selection of the sub-tries of a node (entries of the array) is done by the ordering of the ith character of each key, where i is the depth of the node. The root node uses the first character of the key, the direct descendants of the root use the second character, and so on. At any level where the remaining subtrie has only one record, the branching is suspended. A trie of order M is defined by

tr−M−D : [{tr−M−D}$_1^M$] ; [D] ; nil

The basic trie tree, if the underlying alphabet is represented ordered, is a lexicographically ordered tree. The character set is usually the alphabet or the decimal digits or both. Typically the character set has to include a string-terminator character (blank). If a string terminator character is available, tries can store variable length keys. In particular, as we use the smallest prefix of the key which makes the key unique, digital trees are well suited for handling unbounded or semi-infinite keys.

Let C_n and C_n' denote the average number of internal nodes inspected during a successful search and an unsuccessful search respectively. Let N_n denote the number of internal nodes in a trie with n keys, and let $h(n)$ denote its height. The *digital cardinality* will be denoted by m; this is the size of the alphabet and coincides with the dimension of the internal-node arrays.

In all the following formulas, $P(x)$ denotes complicated periodic (or convergent to periodic) functions with average value 0 and very small absolute value. These functions should be ignored for any practical purposes. Although we use $P(x)$ for all such functions, these may be different.

For tries built from random keys, uniformly distributed in $U(0,1)$ (or keys composed of random-uniform digits) we have:

$$E[N_n] = 1 + m^{1-n} \sum_{i=0}^{n} \binom{n}{i} (m-1)^{n-i} E[N_i] \qquad N_0 = N_1 = 0$$

$$= \frac{n}{\ln m}\left(1 + P(\log_m n)\right) + O(1)$$

$$C_n = 1 + m^{1-n} \sum_{i=1}^{n} \binom{n-1}{i-1} (m-1)^{n-i} C_i, \qquad C_0 = C_1 = 0$$

$$= \frac{H_{n-1}}{\ln m} + \frac{1}{2} + P(\log_m n) + O(n^{-1})$$

$$C_n' = 1 + m^{-n} \sum_{i=2}^{n} \binom{n}{i} (m-1)^{n-i} C_i' \qquad C_0' = C_1' = 0$$

$$= \frac{H_n - 1}{\ln m} + \frac{1}{2} + P(\log_m n) + O(n^{-1})$$

where $H_n = \sum_{i=1}^{n} 1/i$ denote the harmonic numbers.

Digital tree (trie) search

```
search( key, t, depth )
typekey key;
trie t;
int depth;

{
if ( t==NULL )    notfound( key );
else    if ( IsData(t) )
                if ( key == t->k ) found( t );
                        else          notfound( key );
else    search( key, t->p[ charac(depth,key) ], depth+1 );
};
```

Digital tree (trie) insertion

```
trie insert( key, t, depth )
typekey key;
trie t;
int depth;

{
int j;
if ( t==NULL )    return( NewDataNode(key) );
else    if ( IsData(t) )
                if (t->k == key)
                        Error /*** Key already in table ***/;
                else {
                        t = insert( t->k, NewIntNode( ), depth );
                        t = insert( key, t, depth );
                        }
else    {j = charac(depth,key);
        t->p[j] = insert( key, t->p[j], depth+1 );
        }
return( t );
};
```

The function *charac(i,key)* returns the ith character of a key. It is expected that the result is an integer in the range $0...m-1$. Both functions use the level indicator *depth* to facilitate the search. The user should call these functions with depth 1; e.g. *search(key,trie,1)*. The function *IsData(t)* tests whether a pointer points to an internal node or to a data node. The functions *NewIntNode* and *NewDataNode* create new nodes of the corresponding types.

In cases where there is no associated value with the key, we can avoid the data records completely with a special terminator (e.g. nil*) which indicates that a string key terminates there. The key, if desired, can be reconstructed from the path in the tree.

There is a very close correspondence between a trie tree and top-down radix sort, as the trie structure reflects the execution pattern of the sort, each node corresponds to one call to the sorting routine.

	$m=2$		
n	$E[N_n]$	C_n	C_n'
10	13.42660	4.58131	3.28307
50	71.13458	6.96212	5.54827
100	143.26928	7.96937	6.54110
500	720.34810	10.29709	8.85727
1000	1441.69617	11.29781	9.85655
5000	7212.47792	13.62031	12.17792
10000	14425.95582	14.62039	13.17785
	$m=10$		
10	4.11539	1.70903	1.26821
50	20.92787	2.43643	2.05685
100	42.60540	2.73549	2.26860
500	210.60300	3.44059	3.05159
1000	427.45740	3.73802	3.26849
5000	2107.33593	4.44100	4.05106
10000	4275.97176	4.73827	4.26847

Table 3.33: Exact results for general tries.

When the cardinality of the alphabet is large and consequently internal nodes are of significant size compared to a record, the trie becomes inefficient in its use of storage. For example, if only two keys reach a given internal node, we have to include a complete internal node which will be mostly under-utilized. In some sense, tries are efficient close to the root where the branching is dense, but inefficient close to the leaves.

3.4.4.1 Hybrid tries

It is for the above reason that tries are usually composed with some other structure to allow for their efficient behaviour at the root but to switch to some other data structure closer to the leaves. All these compositions have the common definition:

$$\mathbf{tr-M-D : [\{tr-M-D\}_1^M] \; ; \; [D] \; ; \; DICT(D) \; ; \; nil}$$

Common compositions are with external buckets ($\mathbf{DICT(D)} \rightarrow \{\mathbf{D}\}_n^b$), called

Bucket tries and with Binary search trees (**DICT(D)→ bt−D−nil**).

For Bucket tries, after the insertion of n random keys uniformly distributed in $U(0,1)$, we have

$$N_n = \frac{n}{b \ln m}\left(1 + P(\log_m n)\right) + O(1)$$

$$C_n = \frac{H_{n-1} - H_{b-1}}{\ln m} + \frac{1}{2} + P(\log_m n) + O(n^{-1})$$

$$C_n' = \frac{H_n - H_b}{\ln m} + \frac{1}{2} + P(\log_m n) + O(n^{-1})$$

The exact formulas for the above quantities are the same as the ones for general tries but with the extended initial condition: $N_0 = N_1 = ... = N_b = 0$. For Bucket binary tries, i.e. when $m = 2$ we have

$$E[h(n)] = (1 + 1/b)\log_2 n + \frac{\gamma}{\ln 2} + \log_2((b+1)!) + P\left((1+1/b)\log_2 n\right) + o(1)$$

Bucket binary tries are used as the collision resolution mechanism for Dynamic hashing.

A different type of Hybrid trie is obtained by implementing the array in the internal nodes with a structure which takes advantage of its possible sparsity. For example, a linked list consisting of links only for non-empty subtries. Almost any technique of those used for economizing storage in B-tree nodes can be applied to the internal nodes in the tries (see 3.4.2).

3.4.4.2 Tries for word-dictionaries

Digital trees seem very appropriate to implement language dictionaries. The most important reason, besides their efficiency, is that tries allow for efficient prefix searching. *Prefix search* is searching for any word which matches a given prefix, e.g. searching for *comput** where the asterisk can be matched by any string.

There are some problems associated with this particular application though: long common prefixes tend to create unnecessary additional levels with very little (maybe unary) branching. For example, the words *computation, computational, computations* will force 11 levels of branching before these words can be separated. If prefix searching is not needed, this problem can be remedied by organizing the scan of the characters of the key in reverse order (as suffixes are shorter and less common than prefixes).

More generally and much better, if we are prepared to lose the lexicographical ordering of the keys, is to consider the function *charac(i,key)* as a hashing function which operates on the key and returns an integer value with a rather uniform distribution. This option may be particularly appealing when the cardinality of the alphabet is large and the usage distribution is uneven (as would be the case for a full ascii set under normal circumstances). In this latter case the hashing function can be applied to the characters individually.

3.4.4.3 Digital search trees

Digital search trees are a particular implementation of tries where a record is stored in each internal node. The hyperrule which defines these trees is

$$\mathbf{dst-M-D : [D, \{dst-M-D\}_1^M] \ ; \ nil}$$

The binary Digital search trees use the same structure as the Binary search trees; the only difference between them is that the selection of subtrees is not based on comparisons with the key in the node, but on bit inspections.

Let C_n and C_n' be the average number of nodes inspected during a successful and an unsuccessful search respectively. Then for Digital search trees constructed from random uniform keys (or keys composed of random digits) we have:

$$N_n = n$$

$$C_n = 1 + \frac{(n-1)m^{1-n}}{n} \sum_{i=1}^{n} \binom{n-2}{i-1} (m-1)^{n-i-1} C_i \qquad C_0 = 0$$

$$C_n = \log_m n + \frac{\gamma - 1}{\ln m} + \frac{3}{2} - \alpha_m + P(\log_m n) + O\left(\frac{\ln n}{n}\right)$$

$$C_n' = (n+1)C_{n+1} - nC_n - 1$$

$$= \log_m n + \frac{\gamma}{\ln m} + \frac{1}{2} - \alpha_m + P(\log_m n) + O(n^{-1})$$

where

$$\alpha_m = \sum_{k \geq 1} \frac{1}{m^k - 1}$$

$$\alpha_2 = 1.60669...$$

The selection of which key is placed in each node is arbitrary among all the keys of its subtree. As the selected key does not affect the branching (other than by not being in the subtree), any choice will give almost equivalent subtrees. This fact leaves room for optimizing the trees. The most common, and possibly the best, strategy is to choose the key with highest probability. This is equivalent to building the tree by inserting keys in descending probability order.

3.4.4.4 Compressed tries

A Compressed trie is a static tree for which the array of pointers at each internal node is represented by one base address and a bit-array. Each bit indicates whether the corresponding entry points to a non-null sub-trie or not. All non-null sub-tries are stored consecutively starting at the base address.

The easiest way of guaranteeing contiguity is by storing the trie as an array of records. The base address is an integer used as an index into the array. The hyperrule which defines the compressed trie is:

	$m=2$		$m=10$	
n	C_n	C'_n	C_n	C'_n
10	3.04816	3.24647	2.19458	1.64068
50	5.06061	5.41239	2.90096	2.32270
100	6.00381	6.39134	3.19015	2.61841
500	8.26909	8.69616	3.89782	3.31913
1000	9.26011	9.69400	4.18865	3.61622
5000	11.57373	12.01420	4.89731	4.31876
10000	12.57250	13.01398	5.18840	4.61600

Table 3.34: Exact results for Digital search trees.

$$\mathbf{tr-M-D : \{int, \{bool\}_1^M\}_1^N}$$

By convention the root of the trie is at location 1. Given an internal node, its ith descendant will be found by adding the base integer plus the count of "1" bits in the array at the left of location i.

Compressed tries have the same complexity measures as the basic tries. Compressed tries achieve a good efficiency in searching and a very compact representation at the cost of being static structures.

3.4.4.5 Patricia tree

A Patricia tree is a particular implementation of a Binary trie. The Patricia tree uses an index at each node to indicate the bit used for that node's branching. By using this index, we can avoid empty subtrees and hence guarantee that every internal node will have non-null descendants, except for the totally empty tree. A Patricia tree is defined by

$$\mathbf{pat-D : [int, pat-D, pat-D]\ ;\ [D] \quad \equiv \quad bt-int-[D]}$$

As a Binary tree, the Patricia tree stores all its data at the external nodes and keeps one integer, the bit-index, in each internal node.

Let C_n be the average number of internal node inspections during a successful search and C'_n for an unsuccessful search. Then for trees constructed from n randomly distributed keys in $U(0,1)$ we have:

$$N_n = n-1$$

$$C_n = 1 + \frac{1}{2^{n-1}-1}\sum_{i=1}^{n-1}\binom{n-1}{i-1}C_i \qquad C_0 = C_1 = 0$$

$$= \log_2 n + \frac{\gamma}{\ln 2} - \frac{1}{2} + P(\log_2 n) + O(n^{-1})$$

$$C'_n = 1 + \frac{1}{2^n-2}\sum_{i=1}^{n-1}\binom{n}{i}C'_i \qquad C'_0 = C'_1 = 0$$

$$= \log_2 n + \frac{\gamma - \ln\pi}{\ln 2} + \frac{1}{2} + P(\log_2 n) + O(n^{-1})$$

Patricia tree search

```
search( key, t )
typekey key;
patricia t;

{
if (IsData(t))
        if ( key == t->k ) found( t );
                else          notfound( key );
else    if ( bit( t->level, key ) == 0 )
                search( key, t->left );
        else    search( key, t->right );
};
```

Patricia tree insertion

```
typekey samplekey( t )
patricia t;
{return( IsData(t) ? t->k : samplekey( t->left ) ); };

patricia insert( key, t )
typekey key;
patricia t;

{typekey s;
patricia p;
int i;
if (t==NULL)        return( NewDataNode(key) );

/* Find any key in the subtree */
s = samplekey( t );

/* find first different bit */
for (i=1; i<=D && bit(i,key)==bit(i,s); i++);
if (i>D)      Error /* Key already in table */;

else if ( IsData(t) || i<t->level ) {
        /* create a new internal node */
        p = NewDataNode( key );
        if ( bit(i,key)==1 ) t = NewIntNode(i,t,p);
                    else            t = NewIntNode(i,p,t);
        }
```

```
else if (bit(t→level,key)==1)      t→right = insert( key, t→right );
                         else      t→left  = insert( key, t→left  );
return( t );
};
```

The function *bit(i,key)* returns the ith bit of a key. The functions *IsData, NewIntNode* and *NewDataNode* have the same functionality as the ones for tries.

Some implementations keep the number of bits skipped between the bit inspected by a node and the bit inspected by its parent, instead of the bit index. This approach may save some space, but complicates the calling sequence and the algorithms.

Patricia trees are a practical and efficient solution for handling variable length, or very long, keys, particularly well suited for text searching. Note that the problem generated by very long prefixes virtually disappears for Patricia trees.

Given a set of keys, the shape of the tree is determined, so there cannot be any conformation or reorganization algorithm.

n	C_n	C_n'
10	3.58131	3.07425
50	5.96212	5.33950
100	6.96937	6.33232
500	9.29709	8.64847
1000	10.29781	9.64775
5000	12.62031	11.96910
10000	13.62039	12.96903

Table 3.35: Exact results for Patricia trees.

In summary, Digital trees provide a convenient implementation for several database applications. The most important reasons are:

(i) short searching time (successful or unsuccessful);
(ii) allow searching on very long or unbounded keys very efficiently;
(iii) flexibility, as they allow composition with many other structures;
(iv) allow search of interleaved keys and hence they are amenable to multidimensional search.

General References:
[Burkhard, 76], [Burkhard, 76], [Comer *et al.*, 77], [Comer, 79], [Comer, 81], [Flajolet *et al.*, 82], [Flajolet, 81], [Fredkin, 60], [Gotlieb *et al.*, 78], [Horowitz *et al.*, 76], [Knott, 82], [Knuth, 73], [Litwin, 81], [Lomet, 81], [Maly, 76], [Mehlhorn, 79], [Mehlhorn, 81], [Miyakawa *et al.*, 77], [Nicklas *et al.*, 77], [Orenstein, 82], [Regnier, 81], [Tamminen, 81], [Trabb Pardo, 78].

3.5 Multidimensional search

The algorithms which allow non-atomic search keys, or keys composed of several sub-keys, are called multidimensional search algorithms.

Any searching algorithm could, in principle, deal with composite keys, just by considering the composed key as a single block. For this reason only those search algorithms which treat the sub-keys individually are called multidimensional search algorithms. In particular, the most important property of multidimensional search is to allow searching when only some of the sub-keys are specified. This problem is called *partial-match searching* or *partial-match retrieval*. Retrieval on ranges of sub-keys also requires special multidimensional searching algorithms.

Partial-match queries can be treated as a special case of range queries; for a specified sub-key, the range is defined by a single value (upper bound = lower bound), and for an unspecified key the range is infinite (or sufficiently large to include all keys).

General References:
[Bentley *et al.*, 79], [Bentley *et al.*, 79], [Bentley *et al.*, 80], [Bentley, 80], [Bolour, 79], [Dobkin *et al.*, 74], [Dobkin *et al.*, 76], [Guting *et al.*, 80], [Guting *et al.*, 81], [Hirschberg, 80], [Lee *et al.*, 77], [Lee *et al.*, 80], [Lum, 70], [Orenstein, 82], [Ouksel *et al.*, 81], [Raghavan *et al.*, 77], [Saxe, 79], [Tropf *et al.*, 81], [Willard, 82].

3.5.1 Quad trees

A Quad search tree is an extension of the concept of Binary tree search in which every node in the tree has 2^k descendants. While searching for a $k-$dimensional key, the corresponding descendant is selected based on the result of k comparisons.

The descendants of a Quad tree node are numbered from 0 to 2^k-1. Let $b_0b_1...b_{k-1}$ be the binary representation of a descendant number. If b_i is 1 then the ith sub-keys in the descendant subtree are all larger than the ith key at the node; if $b_i=0$ the sub-keys are less or equal. For example, in two dimensions, say x and y, descendant $2 = 10_2$ contains the south-east sector of the plane.

Quad tree search

```
search( key, t )
typekey key[K];
tree t;
{
int i, indx, noteq;
while( t ≠ NULL ) {
        indx = noteq = 0;
        for (i=0; i<K; i++) {
                indx <≤ 1;
                if ( key[i] > t→k[i] ) indx++;
```

```
                if ( key[i] ≠ t→k[i] ) noteq++;
                }
        if ( noteq )  t = t→p[indx];
                else  { found( t ); return; }
        }
notfound( key );
};
```

Quad tree insertion

```
tree    insert( key, t )
typekey key[ ];
tree t;
{
int i, indx, noteq;
if ( t==NULL )    t = CreateNode( key );

else {  indx = noteq = 0;
        for (i=0; i<K; i++) {
                indx <≤ 1;
                if ( key[i] > t→k[i] ) indx++;
                if ( key[i] ≠ t→k[i] ) noteq++;
                }
        if ( noteq )  t→p[indx] = insert( key, t→p[indx] );
        else    Error; /*** Key already in table ***/
        }
return( t );
};
```

There are no efficient or simple methods for performing "rotations" in Quad trees. Consequently it is difficult to maintain a Quad tree balanced.

There are no simple methods for performing deletions either. The best method for deletions is to mark the nodes as deleted, and reconstruct the tree whenever too many nodes have been deleted.

Quad trees have been successfully used to represent data associated with planar coordinates: maps, graphics, bit displays, etc.

Quad trees with dimension 3 or higher become excessively expensive in terms of storage used by pointers. A Quad tree has $(2^k-1)n+1$ null pointers.

Table 3.36 displays simulation results on randomly generated Quad trees. C_n denotes the average successful search and $E[h(n)]$ the average height of a Quad tree with n nodes.

	C_n	
n	$k=2$	$k=3$
5	2.23531 ± 0.00066	2.09322 ± 0.00056
10	2.8431 ± 0.0011	2.53941 ± 0.00091
50	4.3576 ± 0.0030	3.5913 ± 0.0022
100	5.0345 ± 0.0043	4.0502 ± 0.0032
500	6.6222 ± 0.0098	5.1213 ± 0.0072
1000	7.314 ± 0.014	5.576 ± 0.010
5000	8.918 ± 0.031	6.640 ± 0.022
10000	9.599 ± 0.042	7.093 ± 0.031
	$E[h(n)]$	
n	$k=2$	$k=3$
5	3.2844 ± 0.0015	2.9728 ± 0.0014
10	4.4131 ± 0.0027	3.7820 ± 0.0024
50	7.2985 ± 0.0079	5.8197 ± 0.0062
100	8.611 ± 0.012	6.7257 ± 0.0090
500	11.739 ± 0.029	8.874 ± 0.021
1000	13.107 ± 0.042	9.791 ± 0.031
5000	16.350 ± 0.099	11.980 ± 0.072
10000	17.69 ± 0.14	12.96 ± 0.11

Table 3.36: Simulation results for Quad search trees of 2 and 3 dimensions

Under the arbitrary assumption that

$$C_n = \alpha \log_2 n + \beta$$

for $n \geq 100$, we find that

$$\alpha = 0.686 \pm 0.003 \qquad \beta = 0.476 \pm 0.020 \qquad \textit{for } k=2$$

and

$$\alpha = 0.459 \pm 0.002 \qquad \beta = 0.997 \pm 0.015 \qquad \textit{for } k=3$$

References:
[Bentley *et al.*, 75], [Finkel *et al.*, 74], [Flajolet *et al.*, 83], [Gargantini, 82], [Overmars *et al.*, 82], [Samet, 80], [Samet, 83].

3.5.2 K-dimensional trees

A k-d tree is a binary tree which stores k-dimensional keys in its nodes. The sub-keys are used to direct the searching in the same way they are used in a binary search tree. The only difference is that the sub-keys are used cyclically, one sub-key per level. In our algorithm we use the first sub-key at the root, the second sub-key for the direct descendants of the root, and so on.

For k-d trees built from random insertions, the complexity measures are the same as for Binary search trees:

$$E[A_n] = C_n = 2(1+1/n)H_n - 3 \approx 1.3863 \log_2 n - 1.8456,$$

$$\sigma^2(A_n) = (2+10/n)H_n - 4(1+1/n)(H_n^2/n + H_n^{(2)}) + 4$$

$$\approx 1.3863 \log_2 n - 1.4253,$$

$$E[A_n'] = C_n' = 2H_{n+1} - 2 \approx 1.3863 \log_2 n - 0.8456,$$

$$\sigma^2(A_n') = 2H_{n+1} - 4H_{n+1}^{(2)} + 2 \approx 1.3863 \log_2 n - 3.4253,$$

K-d tree search

```
search( key, t )
typekey key[K];
tree t;
{
int indx, i;
for ( indx=0; t ≠ NULL; indx=(indx+1)%K ) {
        for ( i=0; i<K && key[i]==t→k[i]; i++ );
        if ( i==K )  { found( t ); return; }
        if ( key[indx] > t→k[indx] ) t = t→right;
                else              t = t→left;
        }
notfound( key );
};
```

K-d tree insertion

```
tree    insert( key, t, indx )
typekey key[];
tree t;
int indx;
{
int i;
if ( t==NULL )     t = CreateNode( key );

else {  for ( i=0; i<K && key[i]==t→k[i]; i++ );
        if ( i==K )  Error /*** Key already in table ***/;
        else if ( key[indx] > t→k[indx] )
                t→right = insert( key, t→right, (indx+1)%K );
        else    t→left  = insert( key, t→left, (indx+1)%K );
        }
return( t );
};
```

For a k-d tree grown from random keys, a partial-match query which involves p of the k sub-keys will require

$$E[C_n] = O(n^\lambda)$$

where λ is the only positive root of

$$(2+\lambda)^p(1+\lambda)^{k-p} = 2^k$$

The constant which multiplies the n^λ term depends on which sub-keys are used in the partial-match query. This constant is lowest when the sub-keys used for the search are the first sub-keys of the key.

	λ		
k	$p=1$	$p=2$	$p=3$
2	0.56155		
3	0.71618	0.39485	
4	0.78995	0.56155	0.30555

Table 3.37: Order of magnitude of partial-match queries in k-d trees

K-d trees allow range searches; the following algorithm searches a k-d tree for values contained between *lowk* and *uppk*. The function *found*() is called for each value in the tree within the range.

Range search in k-d trees

```
rsearch( lowk, uppk, t, lev )
typekey lowk[], uppk[];
tree t;
int lev;

{int j;
if ( t==NULL )     return;
if ( lowk[lev] ≤ t→k[lev] ) rsearch( lowk, uppk, t→left, lev+1 );

for ( j=0; j<K && lowk[j]≤t→k[j] && uppk[j]≥t→k[j]; j++ );
if ( j==K ) found( t );

if ( uppk[lev] ≥ t→k[lev] ) rsearch( lowk, uppk, t→right, lev+1 );
};
```

There are no efficient or simple methods for performing "rotations" in k-d trees. Consequently it is difficult to maintain a k-d tree balanced.

There are no efficient or simple methods for performing deletions either. The best method for deletions is to mark the nodes as deleted, and reconstruct the tree whenever too many nodes have been deleted.

It is possible to construct a perfectly balanced k-d tree in $O(n \log n)$ time. This is done by a divide and conquer approach:

Construction of perfectly balanced k-d tree

```
MakeBalTree( S : SetOfKeys; i : integer ) : tree;
var   med : typekey;
      A, B : SetOfKeys;
begin
if S=[ ] then MakeBalTree := nil
else begin
      med := median of S in i^th sub-key;
      A := Keys in S smaller than med;
      B := Keys in S greater than med;
      MakeBalTree := NewNode( med,
                              MakeBalTree( A, i+1 mod K ),
                              MakeBalTree( B, i+1 mod K ) )
      end
end;
```

References:

[Bentley, 75], [Bentley, 79], [Eastman *et al.*, 82], [Eastman, 81], [Flajolet *et al.*, 83], [Hoshi *et al.*, 82], [Overmars *et al.*, 82], [Robinson, 81], [Silva Filho, 79], [Silva Filho, 81].

Chapter 4: SORTING ALGORITHMS

4.1 Techniques for sorting arrays

4.1.1 Bubble sort

The Bubble sort algorithm sorts an array by interchanging adjacent records that are in the wrong order. The algorithm makes repeated passes through the array probing all adjacent pairs until the file is completely in order. Every complete pass sets at least one element into its final location (in an upward pass the maximum is settled, in a downward the minimum). In this way, every pass is at least one element shorter than the previous pass.

Let C_n be the number of comparisons needed to sort a file of size n using the Bubble sort, and let I_n be the number of interchanges performed in the process. Then

$$n-1 \leq C_n \leq \frac{n(n-1)}{2}$$

$$E[C_n] = \frac{n^2 - n\ln n - (\gamma + \ln 2 - 1)n}{2} + O(n^{1/2})$$

$$0 \leq I_n \leq \frac{n(n-1)}{2}$$

$$E[I_n] = \frac{n(n-1)}{4}$$

$$E[passes] = n - \sqrt{\pi n/2} + 1/3 + o(1)$$

The simplest form of the Bubble sort always makes its passes from the top of the array to the bottom.

Bubble sort

```
procedure sort( var r : ArrayToSort; lo, up : integer );

var    i, j : integer;
       tempr : ArrayEntry;
begin
while up>lo do begin
       j := lo;
```

```
        for i:=lo to up-1 do
              if r[i].k > r[i+1].k then begin
                    tempr := r[i];
                    r[i] := r[i+1];
                    r[i+1] := tempr;
                    j := i
                    end;
        up := j
        end
end;
```

A slightly more complicated algorithm passes from the bottom to the top, then makes a return pass from top to bottom.

Bubble sort (double direction)

```
sort( r, lo, up )
ArrayToSort r;
int lo, up;

{int i, j;
while (up>lo)       {
      j = lo;
      for ( i=lo; i<up; i++ )
            if ( r[i].k > r[i+1].k ) {
                  exchange( r, i, i+1 );
                  j = i;}
      up = j;
      for ( i=up; i>lo; i-- )
            if ( r[i].k < r[i-1].k )    {
                  exchange( r, i, i-1 );
                  j = i;}
      lo = j;
      }
}
```

The Bubble sort is a simple sorting algorithm, but inefficient. Its running time is $O(n^2)$, unacceptable even for medium-sized files. Perhaps for very small files its simplicity may justify its use, but the Linear insertion sort is just as simple to code and more efficient to run.

For files with very few elements out of place, the double direction Bubble sort (or cocktail shaker sort) can be very efficient. If only k of the n elements are out of order, the running time of the double direction sort is $O(kn)$.

One advantage of the Bubble sort is that it is stable: records with equal keys remain in the same relative order after the sort as before.

References:
[Dobosiewicz, 80], [Knuth, 73], [Meijer *et al.*, 80].

4.1.2 Linear insertion sort

The Linear insertion sort is one of the simplest sorting algorithms. With a portion of the array already sorted, the remaining records are moved into their proper places one by one. This algorithm uses sequential search to find the final location of each element. Let C_n be the number of comparisons needed to sort an array of size n using Linear insertion sort. Then sorting a randomly ordered file requires

$$E[C_n] = \frac{n(n+3)}{4} - H_n$$

$$\sigma^2(C_n) = \frac{(2n-11)n(n+7)}{72} + 2H_n - H_n^{(2)}$$

Linear insertion sort

```
sort( r, lo, up )
ArrayToSort r;
int lo, up;

{int i, j;
ArrayEntry tempr;
for ( i=up-1; i>=lo; i-- ) {
        tempr = r[i];
        for ( j=i+1; j<=up && (tempr.k>r[j].k); j++ )
                r[j-1] = r[j];
        r[j-1] = tempr;
        }
};
```

If the table can be extended to add one *sentinel* record at its end (a record with the largest possible key), Linear insertion sort will improve its efficiency by having a simpler inner loop.

Linear insertion sort with sentinel

```
sort( r, lo, up )
ArrayToSort r;
int lo, up;
```

```
{int i, j;
ArrayEntry tempr;

r[up+1].k = MaximumKey;
for ( i=up-1; i≥lo; i-- ) {
        tempr = r[i];
        for ( j=i+1; tempr.k>r[j].k; j++ )
                r[j-1] = r[j];
        r[j-1] = tempr;
        }
}
```

The running time for sorting a file of size n with the Linear insertion sort is $O(n^2)$. For this reason, the use of the algorithm is justifiable only for sorting very small files. For files of this size (say $n<10$), however, the Linear insertion sort may be more efficient than algorithms which perform better asymptotically. The main advantage of the algorithm is the simplicity of its code.

Like the Bubble sort, the Linear insertion sort is stable: records with equal keys remain in the same relative order after the sort as before.

A common variation of Linear insertion sort is to do the searching of the final position of each key with Binary search. This variation, called *Binary insertion sort,* reduces the number of comparisons but does not reduce the number of interchanges needed to make space for the inserted key. The total running time still remains $O(n^2)$.

References:
[Dijkstra *et al.*, 82], [Doberkat, 82], [Gotlieb *et al.*, 78], [Horowitz *et al.*, 76], [Knuth, 73], [Melville *et al.*, 80].

4.1.3 Quicksort

Quicksort is a sorting algorithm which uses the divide and conquer technique. To begin each iteration an element is selected from the file. The file is then split into two subfiles, those elements with keys smaller than the selected one and those elements whose keys are larger. In this way, the selected element is placed in its proper final location between the two resulting subfiles. This procedure is repeated recursively on the two subfiles and so on.

Let C_n be the number of comparisons needed to sort a random array of size n, let I_n be the number of interchanges performed in the process (for the present algorithm I_n will be taken as the number of record-assignments), and let $k=\lfloor \log_2 n \rfloor$. Then

$$(n+1)k-2^{k+1}+2 \;\le\; C_n \;\le\; \frac{n(n-1)}{2}$$

$$E[C_n] = n-1+\frac{2}{n}\sum_{k=1}^{n-1} E[C_k] = 2(n+1)H_n - 4n$$

$$\sigma^2(C_n) = n\sqrt{7-2\pi^2/3} + o(n)$$

$$E[I_n] = \frac{n+3}{2} + \frac{2}{n}\sum_{k=1}^{n-1} E[I_k] = (n+1)(H_n - 2/3)$$

Below we present the Pascal code for Quicksort. Note that one of the two recursions of the divide and conquer method has been transformed into a **while** loop, like the transformations for tail recursions.

Quicksort algorithm

```
procedure sort( var r : ArrayToSort; lo, up : integer );

var     i, j : integer;
        tempr : ArrayEntry;
begin
while up>lo do begin
        i := lo;
        j := up;
        tempr := r[lo];
        {*** Split file in two ***}
        while i<j do begin
                while r[j].k > tempr.k do
                        j := j-1;
                r[i] := r[j];
                while (i<j) and (r[i].k<=tempr.k) do
                        i := i+1;
                r[j] := r[i]
                end;
        r[i] := tempr;
        {*** Sort recursively ***}
        sort(r,lo,i-1);
        lo := i+1
        end
end;
```

The above algorithm uses the same technique even for very small files. As it turns out, very small sub-files can be sorted more efficiently with other techniques, e.g. Linear insertion sort. It is relatively simple to build a hybrid algorithm which uses Quicksort for large files and switches to a simpler, more efficient, algorithm for small files.

Composition of Quicksort

```
....
begin
while up-lo > M do begin
        .... body of quicksort; ....
        end;
if up > lo then begin
        .... simpler-sort ....
        end
end;
```

Quicksort is a very popular sorting algorithm; although its worst case is $O(n^2)$, its average performance is excellent.

Unfortunately, this worst case occurs when the given file is in order already, a situation which may well happen in practice. Any portion of the file that is nearly in order will significantly deteriorate Quicksort's efficiency. To compensate for this, small tricks in the code of the algorithm can be used to ensure that these worst cases occur only with exponentially small probability.

It should be noted that for the worst case, Quicksort may also use $O(n)$ levels of recursion. This is undesirable, as it really implies $O(n)$ additional storage. Moreover, most systems will have a limited stack capacity. The above algorithm can be protected to force it to use a $O(\log n)$ stack (see Appendix III). In its present form, it will not use $O(n)$ levels of recursion for a file in increasing order.

Quicksort allows several variations, improvements, or mechanisms to protect from its worst case. Most of these variations rely on different methods for selecting the "splitting" element.

(a) The *standard (Quicksort, Quickersort)* algorithms select the splitting element from a fixed location (as in the algorithm above: the first element of the array). Selecting the element in the middle of the array does not deteriorate the random case and improves significantly for partially ordered files.

(b) The variation called *Samplesort* selects a small sample (e.g. size 3) and determines the median of this sample. The median of the sample is used as a splitting element.

(c) The selection of the splitting element can be replaced by a pair of values which determine the range of the median. As the array is scanned, every time an element falls in between the pair, one of the values is updated to maintain the range as close to the median as possible. At the end of the splitting phase we have two elements in their final locations, dividing the interval.

(d) Arithmetic averages, or any other method which selects a value that is not part of the array, produce algorithms that may loop on equal keys.

Arithmetic operations on keys significantly restrict the applicability of sorting algorithms.

n	$E[C_n]$	$E[I_n]$	n	$E[C_n]$	$E[I_n]$
10	24.437	24.885	50	258.92	195.46
100	647.85	456.59	500	4806.41	3069.20
1000	10985.9	6825.6	5000	70963.3	42147.6
10000	155771.7	91218.5	50000	939723.2	536527.6

Table 4.1: Exact average results for Quicksort

References:
[Aho *et al.*, 74], [Apers, 78], [Frazer *et al.*, 70], [Gotlieb *et al.*, 78], [Hoare, 61], [Hoare, 62], [Horowitz *et al.*, 76], [Knuth, 73], [Knuth, 74], [Loeser, 74], [Motzkin, 83], [Peters *et al.*, 75], [Rohrich, 82], [Scowen, 66], [Sedgewick, 77], [Sedgewick, 77], [Sedgewick, 78], [Singleton, 69], [Standish, 80], [vanEmden, 70], [vanEmden, 70], [Wegner, 82].

4.1.4 Shellsort

Shellsort, (or diminishing increment sort) sorts a file by repetitive application of Linear insertion sort. For these iterations the file is seen as a collection of d files interlaced, i.e. the first file is the one in locations $1, d+1, 2d+1, \ldots$, the second in locations $2, d+2, 2d+2, \ldots$, and so on. Linear insertion sort is applied to each of these files for several values of d. For example d may take the values in the sequence $\{n/3, n/9, \ldots, 1\}$. It is crucial that the sequence of increment values ends with 1 (simple linear insertion) to guarantee that the file is sorted.

Different sequences of increments give different performances for the algorithm.

Let C_n be the number of comparisons and I_n the number of interchanges used by Shellsort to sort n numbers.

For $d = \{h, k, 1\}$

$$E[I_n] = \frac{n^2}{4h} + \frac{\sqrt{\pi}}{8}\left(\frac{h^{1/2}}{k} - \frac{ch^{-1/2}}{k} + \frac{c-1}{\sqrt{c}}\right) n^{3/2} + O(n)$$

where $c = gcd(h, k)$.

For $d = \{2^k - 1, 2^{k-1} - 1, \ldots 7, 3, 1\}$

$$E[I_n] = O(n^{3/2})$$

For $d = \{2^k, 2^{k-1}, \ldots, 8, 4, 2, 1\}$ and $n = 2^k$,

$$E[I_n] = \frac{n}{16} \sum_{i=1}^{\log_2 n} \frac{\Gamma(2^{i-1})}{2^i \Gamma(2^i)} \sum_{r=1}^{2^{i-1}} r(r+3)2^r \frac{\Gamma(2^i - r + 1)}{\Gamma(2^{i-1} - r + 1)}$$

$$= 0.534885\ldots n\sqrt{n} - 0.4387\ldots n - 0.097\ldots\sqrt{n} + O(1)$$

$$E[C_n] = E[I_n] + n \log_2 n - \frac{3(n-1)}{2}$$

For $d = \{4^{k+1} + 3 2^j + 1, ..., 77, 23, 8, 1\}$

$$E[I_n] = O(n^{4/3})$$

For $d = \{2^p 3^q, ..., 9, 8, 6, 4, 3, 2, 1\}$

$$E[I_n] = O(n(\log n)^2)$$

For $d = \{\alpha^p(\alpha-1)^q, \ldots, \alpha, \alpha-1\}$ when $\alpha = 2^{\sqrt{\log_2 n}}$

$$E[I_n] = O(n^{1+(2+\varepsilon)/\sqrt{\log_2 n}})$$

for any $\varepsilon > 0$.

The version we present here is closer to the original algorithm suggested by Shell; the increments are $\lfloor n\alpha \rfloor$, $\lfloor \lfloor n\alpha \rfloor \alpha \rfloor, \ldots$. Extensive experiments indicate that the sequence defined by $\alpha = 0.45454 < 5/11$ performs significantly better than other sequences. The easiest way to compute $\lfloor 0.45454n \rfloor$ is by $(5*n-1)/11$ using integer arithmetic. Note that if $\alpha < 1/2$, some sequences will not terminate in 1, but in 0; this has to be corrected as a special case.

Shellsort

```
sort( r, lo, up )
ArrayToSort r;
int lo, up;

{int d, i, j;
ArrayEntry tempr;
for ( d=up-lo+1; d>1; ) {
        if (d<5)      d = 1;
           else d = (5*d-1)/11;
        /*** Do linear insertion sort in steps size d ***/
        for ( i=up-d; i>=lo; i-- ) {
                tempr = r[i];
                for ( j=i+d; j<=up && (tempr.k>r[j].k); j+=d )
                        r[j-d] = r[j];
                r[j-d] = tempr;
                }
        }
}
```

In the following table we present simulation results of the sequence of increments $d = \{\frac{3^k-1}{2}, ..., 40, 13, 4, 1\}$ $(d_{k+1} = 3d_k + 1)$ and for the sequence $d = \{\lfloor n\alpha \rfloor, ..., 1\}$ $(\alpha = 0.45454)$.

	$d_{k+1} = 3d_k + 1$		$\alpha = 0.45454$	
n	$E[C_n]$	$E[I_n]$	$E[C_n]$	$E[I_n]$
10	25.498 ± 0.073	14.106 ± 0.084	25.5113 ± 0.0079	14.1277 ± 0.0091
50	287.27 ± 0.45	164.36 ± 0.46	292.73 ± 0.14	151.46 ± 0.14
100	731.73 ± 0.82	432.44 ± 0.84	738.39 ± 0.27	365.78 ± 0.28
500	5862.7 ± 4.6	3609.4 ± 4.6	5672.8 ± 2.2	2831.4 ± 2.2
1000	$13917. \pm 44.$	$8898. \pm 44.$	13228.1 ± 5.4	6552.9 ± 5.5
5000	$101323. \pm 419.$	$68403. \pm 419.$	$89342. \pm 54.$	$46003. \pm 54.$
10000	$234254. \pm 1397.$	$163355. \pm 1398.$	$194047. \pm 91.$	$97389. \pm 92.$
50000	$1667859. \pm 34116.$	$1235041. \pm 34116.$	$1203412. \pm 598.$	$620169. \pm 602.$

Table 4.2: Simulation results for Shellsort

The simulation results indicate that the performance of both algorithms is rather discontinuous in the size of the file. Consequently, any approximation formula is applicable only in the computed range and will not reflect any discontinuities. For the above simulations, selecting the results with $n \geq 500$ we find the empirical formulas:

$$E[I_n] \approx 1.75n^{1.19} \quad \text{for } \alpha = 0.45454$$

$$\approx 0.48\, n \ln(n) \ln(\ln n)$$

$$E[I_n] \approx 1.29n^{1.28} \quad \text{for } d_{k+1} = 3d_k + 1$$

Shellsort is not a stable sorting algorithm since equal keys may not preserve their relative ordering.

Shellsort seems a very attractive algorithm for internal sorting. Its coding is straightforward and usually results in a very short programme. It does not have a bad worst case and, furthermore, it does less work when the file is partially ordered. These arguments make it a good choice for a library sorting routine.

References:
[Boothroyd, 63], [Erkio, 80], [Espelid, 73], [Ghoshdastidar *et al.*, 75], [Incerpi *et al.*, 83], [Knuth, 73], [Sedgewick, to app.], [Shell, 59], [Yao, 80].

4.1.5 Heapsort

Heapsort (or *Treesort* III) is a sorting algorithm that sorts by building a priority queue and then repeatedly extracting the maximum of the queue until it is empty. The priority queue used is a *heap* that shares the space in the array to be sorted. The heap is constructed using all the elements in the array and is located in the lower part of the array. The sorted array is constructed from top to bottom using the locations vacated by the heap as it shrinks. Consequently we organize the priority queue to extract the maximum element.

$$C_n \leq 2n \lfloor \log_2 n \rfloor + 3n$$

$$I_n \leq n \lfloor \log_2 n \rfloor + 2.5n$$

The complexity results for the heap-creation phase can be found in section 5.1.3.

Heapsort

```
procedure sort( var r : ArrayToSort; lo, up : integer );

var     i : integer;
        tempr : ArrayEntry;
begin
{*** construct heap ***}
for i := (up div 2) downto 2 do    siftup(r,i,up);
{*** repeatedly extract maximum ***}
for i := up downto 2 do begin
        siftup(r,1,i);
        tempr := r[1];
        r[1] := r[i];
        r[i] := tempr
        end
end;
```

The above algorithm uses the function *siftup* (defined in section 5.1.3). A call to *siftup(r,i,n)* constructs a sub-heap in the array r at location i of size n assuming that there are sub-heaps rooted at $2i$ and $2i+1$. Although the above procedure accepts the parameter lo for conformity with other sorting routines, Heapsort assumes that $lo=1$.

Heapsort is not a stable sorting algorithm since equal keys may be transposed.

Heapsort is guaranteed to execute in $O(n\log n)$ time even in the worst case. Heapsort does not benefit from a sorted array, nor is its efficiency significantly affected by any initial ordering. As indicated by simulation, its running time has a very small variance.

This algorithm does not use any extra storage or require languages supporting recursion. Although its average performance is not as good as some other sorting algorithms, the advantages noted indicate that Heapsort is an excellent choice for an internal sorting algorithm.

Heapsort can be modified to take advantage of a partially ordered table. This variation is called *Smoothsort*, and has an $O(n)$ performance for an ordered table and an $O(n \log n)$ performance for the worst case.

Table 4.3 shows simulation results on the total number of comparisons used by Heapsort (C_n) and the total number of interchanges (I_n).

n	$E[C_n]$	$\sigma^2(C_n)$	$E[I_n]$
5	8.86667	1.1475	10.95
10	38.6269 ± 0.0054	3.857 ± 0.015	26.6898 ± 0.0052
50	414.729 ± 0.038	36.68 ± 0.33	241.980 ± 0.031
100	1027.687 ± 0.079	81.2 ± 1.0	581.61 ± 0.065
500	7426.05 ± 0.41	$435. \pm 12.$	4042.34 ± 0.34
1000	16851.81 ± 0.81	$847. \pm 33.$	9081.01 ± 0.67
5000	107685.8 ± 4.2	$4593. \pm 407.$	57104.2 ± 3.4
10000	235371.9 ± 8.1	$8624. \pm 1078.$	124203.2 ± 6.7
50000	$1409794. \pm 40.$	$42214. \pm 9859.$	$737471. \pm 32.$

Table 4.3: Simulation results for Heapsort

The following are approximate formulas computed from the simulation results.

$$E[C_n] \approx 2n\log_2 n - 3.0250n$$

$$\sigma^2(C_n) \approx 0.868n$$

$$E[I_n] \approx n\log_2 n - 0.8610n$$

References:
[Aho *et al.*, 74], [Dijkstra *et al.*, 82], [Dijkstra, 82], [Doberkat, 80], [Floyd, 64], [Hertel, 83], [Horowitz *et al.*, 76], [Knuth, 73], [Standish, 80], [Williams, 64].

4.1.6 Interpolation sort

This sorting algorithm is similar in concept to the Bucket sort. An interpolation function is used to estimate where records should appear in the file. Records with the same interpolation address are grouped together in contiguous locations in the array and later Linear insertion sorted. The main difference between this algorithm and the Bucket sort is that the Interpolation sort is implemented in an array, using only one auxiliary index array and with no pointers.

Let C_n be the number of comparisons needed to sort an array of size n using the Interpolation sort, and let F_n be the total number of interpolation function evaluations made in the process. Then

$$F = 2n$$

$$n-1 \le C_n \le \frac{n(n-1)}{2}$$

$$E[C_n] = \frac{5(n-1)}{4}$$

$$\sigma^2(C_n) = \frac{(20n-13)(n-1)}{72n}$$

The algorithm below uses the interpolation function *phi(key,lo,up)* to sort records of the array. A good interpolation formula for uniformly distributed keys is

General interpolation formula

```
phi( key, lo, up)
typekey key;
int lo, up;

{int i;
i = (key-MinKey) * (up-lo+1) / (MaxKey-MinKey) + lo;
return( i>up ? up : i<lo ? lo : i );
};
```

Note that if the above multiplication is done with integers, this operation is likely to cause overflow.

The array *iwk* is an auxiliary array with the same dimensions as the array to be sorted and is used to store the indices to the working array.

The array *iwk* does not need to be as big as the array to be sorted. If we make it smaller, the total number of comparisons during the final Linear insertion phase will increase. In particular, if *iwk* has m entries and $m \le n$ then

$$E[C_n] = 2n - m - 1 + \frac{n(n-1)}{4m}$$

Interpolation sort

```
sort( r, lo, up )
ArrayToSort r;
int lo, up;

{ArrayIndices iwk;
ArrayToSort out;
ArrayEntry tempr;
int i, j;

for (i=lo; i<=up; i++)    iwk[i] = 0;
for (i=lo; i<=up; i++)    iwk[ phi(r[i].k,lo,up) ]++;
for (i=lo; i<up;  i++)    iwk[i+1] += iwk[i];
for (i=up; i>=lo; i--)    out[ iwk[ phi(r[i].k,lo,up) ]-- ] = r[i];
for (i=lo; i<=up; i++)    r[i] = out[i];
for ( i=up-1; i>=lo; i-- ) {
```

```
            tempr = r[i];
            for ( j=i+1; j≤up && (tempr.k>r[j].k); j++ )
                  r[j-1] = r[j];
            r[j-1] = tempr;
            }
      };
```

The above implementation uses the array *out* to copy the sorted elements. This array can be avoided completely if we can add a flag to each location indicating whether the record has been moved or not.

Because the standard deviation of C_n is $\approx 0.53n^{1/2}$, the total number of comparisons used by the Interpolation sort is very stable around its average.

One of the restrictions of the Interpolation sort is that it can only be used when records have numerical keys which can be handled by the interpolation function. Even in this case, if the distribution of the record key values departs significantly from the uniform distribution, it may mean a dramatic difference in running time. If, however, the key distribution is suitable and we can afford the extra storage required, the Interpolation sort is remarkably fast, with a running time of $O(n)$.

The above implementation of Interpolation sort is stable since equal keys are not transposed.

References:
[Ducoin, 79], [Ehrlich, 81], [Flores, 60], [Gamzon *et al.*, 69], [Gonnet, to app.], [Isaac *et al.*, 56], [Jones, 70], [Kronmal *et al.*, 65], [Tarter *et al.*, 66].

4.1.7 Linear probing sort

This is an Interpolation sort based on a collision resolution technique similar to that of Linear probing hashing. Each key is interpolated into one of the first m positions in an array. (Note that m will be taken to be greater than n unlike most other Interpolation sort methods.) If a collision arises, then the smaller element takes the location in question and the larger element moves forward to the next location, and the process repeats until we find an empty location. (This may, ultimately, cause elements to overflow beyond position m.) After insertion of all elements, a single pass through the array compresses the file to the first n locations. The sorting process can be described as creating a table with Linear probing hashing, using an interpolation function as a hashing function and using the technique of ordered hashing.

Let the size of our table be $m+w$; we will use the first m locations to interpolate the keys and the last w locations as an overflow area. We will let n denote the total number of keys to be sorted and $\alpha = n/m$ be the load factor. Let C_n be the number of comparisons needed to sort the n keys using the Linear probing sort, and let F_n be the total number of interpolation function evaluations performed in the process. Then

$$F_n = n$$

$$E[C_n] = \frac{n}{2}\left(\frac{2-\alpha}{1-\alpha}\right) - \frac{\alpha(\alpha^4-4\alpha^3+6\alpha^2+6)}{12(1-\alpha)^3} + O(m^{-1})$$

Let W_n be the number of keys in the overflow section beyond the location m in the table. We have

$$E[W_n] = \frac{1}{2}\sum_{i=2}^{n}\frac{n^{\underline{i}}}{m^i} = \frac{\alpha^2}{2(1-\alpha)} - \frac{\alpha}{2(1-\alpha)^3 m} + O(m^{-2})$$

where $n^{\underline{i}} = n(n-1)\cdots(n-i+1)$ denotes the descending factorial

$$E[W_m] = \sqrt{m\pi/8} + 2/3 + O(m^{-1/2})$$

$$\sigma^2(W_n) = \frac{6\alpha^2-2\alpha^3-\alpha^4}{12(1-\alpha)^2} + O(m^{-1})$$

$$\sigma^2(W_m) = \frac{(4-\pi)m}{8} + \frac{1}{9} - \frac{\pi}{48} + O(m^{-1/2})$$

$$Pr\{W_n > k\} \le (1-\alpha)\sum_{j\ge 0}\frac{e^{-j\alpha}(j\alpha)^{j+k+1}}{(j+k+1)!}$$

$$\ln\left(\Pr\{W_n > k\}\right) \approx -2k(1-\alpha)$$

Below we describe the Linear probing sort using the interpolation function *phi*(*key*,*lo*,*up*). This sorting function depends on two additional global parameters: m, which is the size of the interpolation area, and *UppBoundr*, which is the upper bound of the input array (*UppBoundr* $\ge m+w$).

Linear probing sort

```
procedure sort( var r : ArrayToSort; lo, up : integer );

var     i, j, uppr : integer;
        r1 : ArrayToSort;
begin
uppr := up + (UppBoundr-up) * 3 div 4;
r1 := r;
for j:=lo to UppBoundr do r[j].k := NoKey;
for j:=lo to up do begin
        i := phi(r1[j].k,lo,uppr);
        while r[i].k ≠ NoKey do begin
                if r1[j].k < r[i].k then begin
                        r1[j-1] := r[i];
                        r[i] := r1[j];
                        r1[j] := r1[j-1]
                        end;
                i := i+1;
                if i > UppBoundr then Error
```

```
                end;
            r[i] := r1[j]
            end;
    i := 0;
    for j:=lo to UppBoundr do
            if r[j].k ≠ NoKey then begin
                    i := i+1;
                    r[i] := r[j]
                    end;
    for j:=i+1 to UppBoundr do        r[j].k := NoKey;
    end;
```

With a good interpolation formula, this algorithm can rank among the most efficient Interpolation sort algorithms.

The application of this algorithm to external storage appears to be promising; its performance, however, cannot be improved by using larger buckets. Letting E_n be the number of external accesses required to sort n records, we have

$$E[E_n] = n\left(1+\frac{\alpha}{2b(1-\alpha)}\right)+\frac{m}{b}$$

Below, table 4.4 gives the efficiency measures for two table sizes with various load factors. I_n denotes the number of interchanges performed due to collisions while building the table.

	$m=100$			$m=5000$		
α	$E[C_n]$	$E[W_n]$	$E[I_n]$	$E[C_n]$	$E[W_n]$	$E[I_n]$
50%	72.908	0.23173	13.83±0.13	3747.65	0.24960	766.4±5.9
80%	200.697	1.27870	85.57±0.74	11932.90	1.59014	5941.±67.
90%	310.185	2.47237	164.1±1.4	24149.77	3.96477	16130.±273.
95%	399.883	3.62330	233.8±1.9	45518.47	8.39737	36121.±846.
99%	499.136	5.10998	315.8±2.4	118134.0	26.46562	106957.±3475.
100%	528.707	5.60498		169087.0	43.64541	

Table 4.4: Exact and simulation results for Linear probing sort

References:
[Gonnet *et al.*, 81], [Melville *et al.*, 80].

4.1.8 Summary

The following table shows an example of real relative total times for sorting an array with 49998 random elements.

Algorithm	C	Pascal
Quicksort		1.5
Shellsort	2.1	2.2
Heapsort	2.6	3.2
Interpolation sort	1.0	1.2
Linear probing sort	1.0	1.0

Table 4.5: Relative total times for sorting algorithms

General References:
[Aho *et al.*, 83], [Boothroyd, 63], [Borodin *et al.*, 79], [Borodin *et al.*, 82], [Cook *et al.*, 80], [Erkio, 81], [Flores, 61], [Flores, 69], [Friend, 56], [Hibbard, 63], [Horvath, 78], [Knuth, 74], [Lorin, 75], [Manacher, 79], [Martin, 71], [Mehlhorn, 79], [Mehlhorn, 81], [Pohl, 75], [Trabb Pardo, 77], [Wirth, 76].

4.2 Sorting other data structures

The second most popular data structure used to store sorted data is the linked list, or linear list. The corresponding data structure is described by the production:

s−KEY : [KEY, s−KEY] ; nil

A typical Pascal definition of a linked list, containing a key-field *k*, is:

Linked list definition

```
type
    list = ↑rec;
    rec  = record
             k : typekey;
             next : list
           end;
```

Linked lists can be implemented in arrays; in this case a pointer to a record is an integer indexing into the array. The only non-trivial operation when implementing lists in arrays is to reorder the array according to the order given by the list. This is particularly useful for the case of sorting. The following algorithm reorders the array *r* based on the list rooted at *root*.

Reordering of arrays

```
i := 1;
while root ≠ 0 do begin
      tempr := r[root];
```

```
        r[root] := r[i];
        r[i] := tempr;
        r[i].next := root;
        root := tempr.next;
        i := i+1;
        while (root<i) and (root>0) do   root := r[root]↑.next;
        end;
```

General References:
[Aho *et al.*, 83], [Bentley *et al.*, 80], [Borodin *et al.*, 79], [Borodin *et al.*, 82], [Chin *et al.*, 80], [Colin *et al.*, 80], [Flores, 69], [Friend, 56], [Gotlieb *et al.*, 78], [Harper *et al.*, 75], [Munro *et al.*, 76], [Nozaki, 79], [Power, 80], [Sedgewick, 78], [Tanner, 78], [Tarjan, 72], [Wegner, 82], [Wirth, 76].

4.2.1 Merge sort

Merge sort is a natural way of sorting lists by repeated merging of sublists. By counting the total number of records in the list, each merging step can be as balanced as possible. At the deepest level of the recursion, single element lists are merged together to form two element lists and so on.

Let C_n the the total number of comparisons used by Merge sort, then

$$\sum_{i=1}^{n-1} \nu(i) \le C_n \le kn - 2^k + 1$$

where $k = \lceil \log_2 n \rceil$ and $\nu(i)$ is the number of 1's in the binary representation of i.

$$k 2^{k-1} \le C_{2^k}$$

$$E[C_{2^k}] = (k-\alpha)2^k + 2 - \frac{4.2^{-k}}{3} + \frac{8.4^{-k}}{7} - \frac{16.8^{-k}}{15} + O(16^{-k})$$

$$(\log_2 n - \alpha)n + 2 + O(n^{-1}) \le E[C_n] \le (\log_2 n - \beta)n + 2 + O(n^{-1})$$

where $\alpha = 1.26449... = 2 - \sum_{i \ge 0} \frac{1}{2^i(2^i+1)}$ and $\beta = 1.24075....$

Merge sort

```
function sort( var r : list; n : integer ) : list;
var     temp : list;

begin
if r = nil then sort := nil
else    if n>1 then
        sort := merge( sort( r, n div 2 ),
                       sort( r, (n+1) div 2 ))
```

```
else    begin
        temp := r;
        r := r↑.next;
        temp↑.next := nil;
        sort := temp
        end
end;
```

It is assumed that we know the number of elements in the list, which is given as the second parameter. If this number is not known, it can be overestimated without deteriorating significantly the performance of the algorithm.

The function *merge* merges two ordered lists into a single list and is described in section 4.3.1.

If the merging routine is stable, i.e. in the output of *merge(a,b)* equal keys are not transposed and those from the list *a* precede those from the list *b*, Merge sort will be a stable sorting algorithm and equal keys will not be transposed.

Merge sort uses extra storage: the pointers that are associated with the list.

Merge sort can take advantage of partially ordered lists *(Natural merge)* as described in appendix III. For this variation, the algorithm will do a single pass on totally ordered (or reversely ordered) files and will have a smooth transition between $O(n)$ and $O(n \log n)$ complexity for partially ordered files. Merge sort is guaranteed to execute in $O(n \log n)$ even in the worst case.

In view of the above, Merge sort is one of the best alternatives to sorting lists.

Table 4.6 illustrates some exact counts of number of comparisons for Merge sort. The average values are computed for random permutations of the input file.

n	min C_n	$E[C_n]$	max C_n
5	5	7.1667	8
10	15	22.667	25
50	133	221.901	237
100	316	541.841	573
500	2216	3854.58	3989
1000	4932	8707.17	8977
5000	29804	55226.3	56809
10000	64608	120450.7	123617
50000	382512	718184.3	734465

Table 4.6: Number of comparisons used by Merge sort

References:
[Aho *et al.*, 74], [Bron, 72], [Dewar, 74], [Erkio, 80], [Horowitz *et al.*, 76], [Jones, 70], [Knuth, 73], [Peltola *et al.*, 78], [Todd, 78].

4.2.2 Quicksort for lists

A natural way of sorting a list is by the use of the divide and conquer technique. This will produce an algorithm similar to Quicksort; i.e. pick an element of the list (the head of the list), split the remaining list according to elements being smaller or larger than the selected one, sort the two resulting lists recursively, and finally concatenate the lists.

The execution pattern (sizes of subfiles, etc.) of this algorithm is the same as for Quicksort for arrays. Let I_n be the number of times the inner loop is executed to sort a file with n elements. The inner loop involves one or two comparisons and a fixed number of pointer manipulations. Let C_n be the number of comparisons and $k = \lfloor \log_2 n \rfloor$, then

$$(n+1)k - 2^{k+1} + 2 \;\le\; I_n \;\le\; \frac{n(n-1)}{2}$$

$$E[I_n] \;=\; n-1+\frac{2}{n}\sum_{i=1}^{n-1} E[I_i] \;=\; 2(n+1)H_n - 4n$$

$$E[C_n] \;=\; \frac{3I_n}{2}$$

$$\sigma^2(C_n) = n\sqrt{7-2\pi^2/3} + o(n)$$

Quicksort for lists

```
function sort ( r : list ) : list;
var    lowf,lowl, midf,midl, highf,highl : list;

begin
if r = nil then        begin Last := nil;  sort := r end
else   begin
       lowf := nil;  midf := nil;  highf := nil;
       {*** First key becomes splitter ***}
       tailins( r, midf, midl );
       r := r↑.next;
       while r≠nil do begin
             if r↑.k<midf↑.k then        tailins(r,lowf,lowl)
             else if r↑.k=midf↑.k then   tailins(r,midf,midl)
                      else    tailins(r,highf,highl);
             r := r↑.next
             end;
       {*** Assemble resulting list ***}
       if lowf ≠ nil then begin
             lowl↑.next := nil;
```

```
                sort := sort(lowf);
                Last↑.next := midf
                end
         else   sort := midf;
         if highf ≠ nil then highl↑.next := nil;
         midl↑.next := sort(highf);
         if Last = nil then Last := midl
         end
end;
```

This algorithm keeps track of lists by keeping a pair of pointers to each list: one to the head and one to the tail. This is particularly useful for concatenating lists together. The global variable *Last* is used to return a pointer to the last element of a sorted list. The procedure *tailins* inserts a record at the end of a list given by a pair of pointers.

Insert a record at the end of a list

```
procedure tailins ( rec : list; var first,last : list );
begin
if first = nil then   first := rec
               else   last↑.next := rec;
last := rec
end;
```

The worst case, $O(n^2)$ comparisons, happens, among other cases, when we sort a totally ordered or reverse-ordered list.

The above implementation of Quicksort keeps a list of all the elements that are equal to the splitting record. By doing this, and by growing the lists at the tail, Quicksort for lists becomes a stable sorting algorithm, i.e. equal keys are not transposed.

When sorting lists with Quicksort we cannot easily prevent the worst case. Consequently, portions of the list that are already in order will deteriorate the algorithm's performance significantly.

References:
[Motzkin, 81], [Wegner, 82].

4.2.3 Bucket sort

Bucket sort (or *Address-calculation sort)* uses an interpolation formula on the keys to split the records between m buckets. The *buckets* are sets of records, which we implement using lists. After the splitting pass, the records in the first bucket will have smaller keys than the records in the second bucket and so on. The buckets are then sorted recursively and finally all the buckets are concatenated together.

Let I_n denote the number of times that a key is placed into a bucket. This measure counts the number of times the innermost loop is executed. I_n satisfies the recurrence equation:

$$I_n = n + \sum_{i=0}^{n-2} \binom{n-2}{i} \frac{(m-1)^{n-2-i}}{m^{n-2}} \left((m-2)I_i + 2I_{i+1}\right)$$

for fixed m, and $m<n$ the solution of the above is

$$I_n = n \log_m n + n\,Q(\log_m n, m) - \frac{1}{2\ln m} + O(n^{-1})$$

where $Q(x,m)$ is a periodic function in x. For m proportional to n, $n=m\alpha$, (m remains fixed for the recursive calls) then

$$I_n = (2-e^{-\alpha})n + \alpha^2 - 2 + e^{-\alpha}(\alpha^2/2+\alpha+2) + O(n^{-1})$$

For $m=n$, (and m is set equal to n for each recursive call)

$$I_n = n + \sum_{i=0}^{n-2} \binom{n-2}{i} \frac{(1-1/n)^{n-2}}{(n-1)^i} \left((n-2)I_i + 2I_{i+1}\right)$$

$$= 1.76108...n - 0.39125... + O(n^{-1})$$

Bucket sort

```
list sort( s, min, max )
list s;
typekey min, max;

{
int i;
typekey div, maxb[M], minb[M];
list head[M], t;
struct rec aux;
extern list Last;
if (s==NULL)        return(s);
if (max==min)       {
        for (Last=s; Last→next≠NULL; Last = Last→next);
        return( s );
        }
div = (max−min) / M;      /* Find dividing factor */
if (div==0)  div = 1;
for (i=0; i<M; i++)       head[i] = NULL;
/* Place records in buckets */
while (s ≠ NULL)  {
        i = (s→k−min) / div;
```

```
        if (i<0) i = 0;     else  if (i≥M) i = M−1;
        t = s;
        s = s→next;
        t→next = head[i];
        if (head[i]==NULL)        minb[i] = maxb[i] = t→k;
        head[i] = t;
        if ( t→k > maxb[i] )      maxb[i] = t→k;
        if ( t→k < minb[i] )      minb[i] = t→k;
        }
/* sort recursively */
t = &aux;
for (i=0; i<M; i++)     if (head[i]≠NULL) {
        t→next = sort( head[i], minb[i], maxb[i] );
        t = Last;
        }
return(aux.next);
}
```

The above algorithm computes the maximum and minimum key for each bucket. This is necessary and convenient as it allows correct sorting of files containing repeated keys and reduces the execution time. Bucket sort requires two additional parameters, the maximum and minimum key. Since these are recomputed for each pass, any estimates are acceptable; at worst, if the estimate is pessimal, it will force Bucket sort into one additional pass.

The above function sets the global variable *Last* to point to the last record of a sorted list. This allows easy concatenation of the resulting lists.

Bucket sort can be combined with other sorting techniques. If the number of buckets is significant compared to the number of records, most of the sorting work is done during the first pass. Consequently we can use a simpler (but quicker for small files) algorithm to sort the buckets.

Although the worst case for Bucket sort is $O(n^2)$, this can only happen for particular sets of keys and only if the spread in values is $n!$. This is very unlikely. If we can perform arithmetic operations on keys, Bucket sort is probably the most efficient alternative to sorting lists.

References:
[Akl *et al.*, 82], [Cooper *et al.*, 80], [Devroye *et al.*, 81], [Flores, 60], [Isaac *et al.*, 56], [Knuth, 73], [Tarter *et al.*, 66].

4.2.4 Radix sort

Radix *(or distribution)* sort, sorts records by examining the digital decomposition of the key. This algorithm does one pass of the file for each digit in the key. In each pass, the file is split according to the values of the corresponding digit. The sorting can be done top-down or bottom-up depending on the relative order of the splitting and recursive sorting of the subfiles.

If we split, sort recursively and concatenate, the resulting algorithm, which we will call *Top-down Radix sort,* resembles Bucket sort, where instead of computing a "bucket address" the bucket is selected based on a digit of the key.

The *bottom-up Radix sort,* where we sort recursively, split and concatenate, is the most common version of Radix sort. This method was at one time very popular in the data-processing field as it is the best method for sorting punched cards.

There is a close correspondence between the top-down Radix sort and Digital trees or tries. The number of times a given record is analyzed corresponds to the depth of the record in an equivalent trie tree. The total complexity, i.e. total number of records passed, coincides with the internal path in an equivalent trie. These results can be found in section 3.4.4.

For the bottom-up algorithm, let m be the base of the numbering system, let D be the number of digits in the key and let I_n be the number of times the innermost loop is repeated (number of records passed through). Then

$$I_n = nD$$

It is possible to group several digits together, in which case D and m could vary as long as

$$m^D = K_1$$

(where K_1 is a constant for a given file). Given this constraint, the tradeoffs are simple: the time complexity is linear in D and the additional storage is linear in m.

Bottom-up Radix sort

```
function sort( r : list ) : list;
var     head, tail : array[1..M] of list;
        i, j, h : integer;

begin
for i:=D downto 1 do begin
        for j:=1 to M do head[j] := nil;
        while r ≠ nil do begin
                h := charac( i, r↑.k );
                if head[h]=nil then head[h] := r
                else    tail[h]↑.next := r;
                tail[h] := r;
                r := r↑.next;
                end;
        {*** Concatenate lists ***}
        r := nil;
        for j:=M downto 1 do
                if head[j] ≠ nil then begin
```

```
                    tail[j]↑.next := r;
                    r := head[j]
                    end
        end;
    sort := r
    end;
```

The above sorting algorithm uses the function *charac(i,key)* which returns the ith digit from the key *key*. The top-down Radix sorting function is described in appendix III.

If $D \log m$ is larger than $\log n$ then bottom-up Radix sort is not very efficient. On the other hand, if $D \log m < \log n$ (some keys must be duplicated), Radix sort is an excellent choice.

References:
[Aho *et al.*, 74], [Hibbard, 63], [Knuth, 73], [McCulloch, 82].

4.2.5 Hybrid methods of sorting

Most of the sorting algorithms described so far are basic in the sense that their building blocks are more primitive operations rather than other sorting algorithms. In this section we describe algorithms which combine two or more sorting algorithms. The basic sortings usually have different properties and advantages and are combined in a way to exploit their most advantageous properties.

4.2.5.1 Recursion termination

This is a general technique which has been described for Quicksort in particular. Many recursive sorting algorithms have good general performance, except that they may do an inordinate amount of work for a file with very few elements (e.g. Quicksort or Bucket sort for two elements).

On the other hand, being efficient for the tail of the recursion is very important for the total complexity of the algorithm.

The general scheme for hybrid recursive sorts is then

Hybrid termination

```
function sort( keys );
begin
if size(keys) > M then
        < ...main sorting algorithm... >
else    simplersort( keys );
end;
```

The *simplersort()* part may be just an analysis of 1,2, and 3 elements by brute force or another sorting algorithm which does well for small files. In the latter case, Linear insertion sort is a favourite candidate.

4.2.5.2 Distributive partitioning

Distributive partitioning sort is a composition of a balanced Quicksort with Bucket sort. The file is split by the median element (or an element chosen to be very close to the median, e.g. median of medians) and then the lower and upper elements, separately, are Bucket-sorted. The procedure may be applied recursively, or we may use still another composition for sorting the individual buckets.

The motive for this composition is to profit from the good average performance of Bucket sort, while guaranteeing an $O(n \log n)$ time by splitting the file by the median at each step.

$$C_n = O(n \log n)$$

If the median is too costly to compute we could split the file into two equal-size parts and apply Bucket sort twice. We then sort the buckets recursively and finally merge the two halves. This has the same the same effect as computing the median for the worst case, but it is much more efficient.

4.2.5.3 Non-recursive Bucket sort

When the number of buckets is relatively large, Bucket sort achieves an excellent average performance ($O(n)$). Not only is the time linear, but the constant in the linear term is reasonably small; the first pass does most of the sorting.

However, the $O(n^2)$ worst case is clearly unwanted. A family of hybrid algorithms can be derived from compositions of a single pass of Bucket sorting and a second algorithm. This second algorithm should: (a) sort small files efficiently, as this is what it will do most; (b) have an $O(n \log n)$ worst-case performance, in case Bucket sort hits an "unlucky" distribution.

Again, we could have a double composition, one algorithm good for case (a) and one good for case (b). For example we could use Linear insertion sort for less than 10 elements and Heapsort otherwise.

Another alternative is to use Natural merge sort. The worst case for Bucket sort (batches of equal keys) is almost the best case for Natural merge sort.

References:
[Dobosiewicz, 78], [Dobosiewicz, 79], [Huits *et al.*, 79], [Jackowski *et al.*, 79], [Nat, 80], [Peltola *et al.*, 78].

4.2.6 Treesort

A Treesort sorting algorithm sorts by constructing a lexicographical search tree with all the keys. Traversing the tree in an infix order, all the nodes can be output in the desired order. Treesort algorithms are a composition of search tree insertion with infix tree traversal.

The number of comparisons required to sort n records is related to the specific type of search tree. Let C_n be the average number of comparisons in a successful search, then

$$C_n^{sort} = nC_n$$

Almost any of the tree structures described in section 3.4 can be used for this purpose. The following algorithm is based on binary trees.

Binary Treesort

```
tree := nil;
for i:=1 to n do insert( tree, <next-key> );
output_infix( tree );
```

where the function *output_infix* is

Scan binary tree in infix order

```
procedure output_infix( t : tree );
begin
if t ≠ nil then begin
        output_infix( t↑.left );
        output( t↑.key );
        output_infix( t↑.right );
        end
end;
```

These algorithms require two pointers per record and consequently are significantly more expensive than other methods in terms of additional storage. There is one circumstance when this structure is desirable, and that is when the set of records may grow or shrink, and we want to be able to maintain it in order at low cost.

To guarantee an $O(n \log n)$ performance it is best to select some form of balanced tree (AVL, weight balanced, B-trees, etc.).

References:

[Aho *et al.*, 74], [Woodall, 71].

4.3. Merging

A special case of single sorting is to build a sorted file from several sorted files. This process is called *merging* of files and it is treated separately, as it normally requires simpler algorithms.

Merging a small number of files together is easily achieved by repeated use of a function which merges two files at a time. In most cases, an optimal strategy is to repeatedly merge the two smallest files until there is only one file left. For this reason, the merging of two ordered files is the main function which we will analyze in this section. Algorithms for merging large numbers of files are studied in conjunction with external sorting. In particular, the second phases of the Merge sort algorithms are good merging strategies for many files.

A *stable merging* algorithm is one which preserves the relative orderings of equal elements from each of the sequences. The concept of stability can be extended to enforce that equal elements between sequences will maintain a consistent ordering; this is called *full stability*.

General References:
[Fabri, 77], [Floyd *et al.*, 73], [Harper *et al.*, 75], [Hyafil *et al.*, 74], [Nat, 79], [Schlumberger *et al.*, 73], [Sedgewick, 78], [Tanner, 78], [Yao *et al.*, 76].

4.3.1. List merging

Merging two ordered lists is achieved by repeatedly comparing the top elements and moving the smallest-key one to the output list.

Assuming that all the possible orderings of the lists are equally likely, then:

$$E[C_{n_a,n_b}] = n_a + n_b - \frac{n_a}{n_b+1} - \frac{n_b}{n_a+1}$$

$$\sigma^2(C_{n_a,n_b}) = \frac{n_b(2n_b+n_a)}{(n_a+1)(n_a+2)} + \frac{n_a(2n_a+n_b)}{(n_b+1)(n_b+2)} - \left(\frac{n_a}{n_b+1} + \frac{n_b}{n_a+1}\right)^2$$

$$C^{\max}_{n_a,n_b} = n_a + n_b - 1$$

List merging (Pascal)

```
function merge ( a, b : list ) : list;

var   first, last, temp : list;
begin
first := nil;
while b ≠ nil do
        if a = nil then       begin a := b;  b := nil end
```

```
else    begin
        if b↑.k > a↑.k then
                begin temp := a; a := a↑.next end
        else    begin temp := b; b := b↑.next end;
        temp↑.next := nil;
        tailins( temp, first, last )
        end;
tailins( a, first, last );
merge := first
end;
```

The above function uses the procedure *tailins* which inserts a node into a list defined by its first and last pointers. Such a procedure is useful in general for working with lists and is described in section 4.2.

The above algorithm is stable but not fully stable.

References:
[Horowitz *et al.*, 76], [Knuth, 73].

4.3.2. Array merging

Array merging is a simple operation if enough additional space is available. For example, merging two arrays with additional space amounting to the smallest of the arrays can be accomplished in n_a+n_b-1 comparisons. The next algorithm merges arrays a and b of size n_a and n_b respectively into the array a.

Merging of arrays

```
merge( a, b, na, nb )
RecordArray a, b;
int na, nb;

{ /*** Merge array b (0...nb-1) into array a (0...na-1) ***/
while ( nb > 0 )
        if ( na<=0 || a[na-1].k < b[nb-1].k )
                        { nb--; a[na+nb] = b[nb]; }
                else    { na--; a[na+nb] = a[na]; }
};
```

There are several algorithms to merge arrays with little or no additional storage. However, these are quite complex. The problem can be slightly rephrased, and in that case is usually referred to as *in-place merging:* given an array a which contains two sequences of ordered elements, one in locations

$1...n_a$ and the other in locations $n_a+1...n_a+n_b$, merge them into one sequence using only m units of additional storage.

Most of these algorithms, although asymptotically better, will not compete with an in-place sorting method for practical purposes. In particular, Shellsort (section 4.1.4) will do less work for the merging of two sequences than for sorting a random array, and is thus recommended.

Comparisons	Extra space	Stable	Reference
$O(n)$	$O(1)$	No	Kronrod
$O(n)$	$O(\log n)$	Yes	Horvarth
$O(n)$	$O(1)$	Yes	Trabb
$O(kn)$	$O(n^{1/k})$	Yes	Wong
$O(n_b \log(n_a/n_b+1))$	$O(\log n_b)$	Yes	Dudzinski & Dydek

Table 4.7: Characteristics of in-place merging algorithms

Table 4.7 lists the properties and references for some in-place merging algorithms. n_a and n_b denote the sizes of the two arrays to be merged, $n_a+n_b=n$, and without loss of generality we assume $n_a \ge n_b$.

References:
[Dudzinski *et al.*, 81], [Horvath, 74], [Horvath, 78], [Knuth, 73], [Kronrod, 69], [Trabb Pardo, 77], [Wong, 81].

4.3.3. Minimal-comparison merging

Let $C^{MM}_{n_a,n_b}$ denote the minimum-maximum, or the minimum worst-case number of comparisons required to merge two files of sizes n_a and n_b. It is known that

$$C^{MM}_{n,1} = \lceil \log_2 n+1 \rceil$$

$$C^{MM}_{n,2} = \lceil \log_2 \frac{7(n+1)}{6} \rceil + \lceil \log_2 \frac{7(n+1)}{17} \rceil$$

$$C^{MM}_{n,3} = \lceil \log_2 \frac{7n+13}{17} \rceil + \lceil \log_2 \frac{7(n+2)}{107} \rceil + \lceil \log_2 \frac{7(n+2)}{43} \rceil + 5 \qquad (n \ge 9)$$

$$C^{MM}_{n_a,n_b} \ge \lceil \log_2 \binom{n_a+n_b}{n_a} \rceil$$

$$C^{MM}_{n_a,n_b} = n_a+n_b-1$$

if $\min(n_a,n_b) \ge 2\,|n_a-n_b|-2$.

The Hwang and Lin merging algorithm, sometimes called *binary merging,* merges two files with an almost optimal number of comparisons. This algorithm is optimal for merging a single element into a sequence, two equal sequences and other cases. Compared to the standard algorithm, it reduces the number of comparisons significantly for files with very different sizes, however the number of movements will not be reduced, and hence this algorithm is mostly of theoretical interest.

The basic idea of binary merging is to compare the first element of the shorter file with the 1st or 2nd or 4th or 8th,... element of the longer file, depending on the ratio of the file sizes. If $n_a \geq n_b$ then we compare the first element of file b with the 2^t element of a, where $t = \lfloor \log_2 n_a / n_b \rfloor$. If the key from file b comes first, then a binary search between $2^t - 1$ elements is required; otherwise 2^t elements of file a are moved ahead. The procedure is repeated until one of the files is exhausted.

In its worst case, Hwang and Lin's algorithm requires

$$C^{HL}_{n_a, n_b} = (t+1)n_b + \lfloor n_b / 2^t \rfloor - 1$$

where $t = \lfloor \log_2 n_a / n_b \rfloor$.

Manacher introduced an improvement to the Hwang and Lin algorithm when $n_a / n_b \geq 8$, which reduces the number of comparisons by $n_b / 12 + O(1)$.

References:
[Christen, 78], [Hwang *et al.*, 71], [Hwang *et al.*, 72], [Hwang, 80], [Knuth, 73], [Manacher, 79], [Schulte Monting, 81], [Stockemyer *et al.*, 80], [Thanh *et al.*, 82].

4.4 External sorting

Sorting files that do not fit in internal memory, and are therefore stored in external memory, requires algorithms which are significantly different from those used for sorting internal files. The main differences are:

(i) the most expensive operation is accessing (or storing) a record;

(ii) the intermediate files may not support direct (or random) access of elements, and even if they do support direct accesses, sequential accesses are more efficient.

Our main measure of complexity is the number of times that the file has been copied, or read and written. A complete copy of the file is called a *pass*.

The algorithms we will describe use the following interface with the file system:

Interface with file system

```
function ReadFile( i : integer ) : record;
procedure WriteFile( i : integer; r : record );
procedure OpenWrite( i : integer );
function OpenRead( i : integer ) : record;
function Eof( i : integer ) : boolean;
ReadDirect( i : integer ) : record;
WriteDirect( i : integer; r : record );
```

In all cases the argument i refers to a unit number, an integer in the range 1...*maxfiles*. The function *Eof(i)* returns the value "*true*" when the last *ReadFile* issued failed. The functions *OpenWrite* and *OpenRead* set the corresponding indicator to the letters "o" (output unit) and "i" (input unit) respectively in the global array *FilStat*. The direct access operations use an integer to select the record to be read/written. These operations use the input file only. Without loss of generality we will assume that the input file is in unit 1, which can be used later for the sorting process. Furthermore, the output file will be placed in any file whose index is returned by the sorting procedure. In the worst case, if this is not desired and cannot be predicted, a single copy is sufficient.

The *external merge sorting* algorithms are the most common algorithms and use two phases: distribution phase and merging phase. During the *distribution phase* or *dispersion phase* the input file is read and sorted into sequences, each sequence as long as possible. These sequences, sometimes called *strings* or *runs*, are distributed among the output files. The *merging phase* merges the ordered sequences together until the entire file is a single sequence; at this point the sorting is completed.

The options available for creating the initial sequences (runs), for distributing them and organizing the merging phase (which files to merge with which, etc.) give rise to many variations of external merge sorting.

The distribution phase's objective is to create as few sequences as possible, and at the same time distribute these sequences in a convenient way to start the merging phase. There are three main methods for constructing the ordered sequences: *replacement selection, natural selection* and *alternating selection.*

4.4.0.1 Replacement selection

The replacement selection algorithm keeps records in an internal buffer of size M. When the buffer is full, the smallest-key record is output and a new record is read. Subsequently, the smallest-key record in the buffer, whose key is larger than or equal to the last written, is output, and a new record is read. When this is no longer possible, i.e. all keys in the buffer are smaller than the last output, a new sequence is initiated.

The expected length of the ith run, denoted by n_i, is

$$E[n_i] = L_i M = 2M + O(M 8^{-i})$$

$$E[n_1] = (e-1)M$$

where the values L_i are given by the generating function

$$L(z) = \sum_{i \ge 0} L_i z^i = \frac{z(1-z)}{e^{z-1} - z} - z$$

The simplest way to manage the buffers is to keep a priority queue with the elements larger than the last output key, and a pool with the others. The following code describes the function *distribute* which uses a heap as a priority queue.

Replacement selection distribution

```
distribute( )

{int i, hbot, s;
typekey lastout;

for (i=0; i<M; i++) {
        Buff[i] = ReadFile( 1 );
        if ( Eof(1) ) break;
        }
i--;

while (i≥0) {
        for (hbot=0; hbot<i; )      insert( ++hbot, Buff );
        /*** Start a new sequence ***/
        s = nextfile();
        while (hbot ≥ 0) {
                lastout = Buff[0].k;
                WriteFile( s, Buff[0] );
                Buff[0] = Buff[hbot];
                siftup( Buff, 0, hbot-1 );
                if ( !Eof(1) ) Buff[hbot] = ReadFile(1);
                if ( Eof(1) )  Buff[hbot--] = Buff[i--];
                else if ( Buff[hbot].k < lastout )  hbot--;
                        else   insert( hbot, Buff );
                }
        }
};
```

The function *nextfile* returns the file number on which the next sequence or run should be placed. The functions *insert* and *siftup* are described in the priority queue section, 5.1.3.

References:
[Dinsmore, 65], [Espelid, 76], [Gassner, 67], [Goetz, 63], [Knuth, 63], [Knuth, 73].

4.4.0.2 Natural selection

Natural selection is a mechanism for producing runs, similar to Replacement selection, which uses a *reservoir* of records to increase the efficiency of the internal buffer. Until the reservoir is full, new records with keys smaller than the last output record are written into the reservoir. Once the reservoir is full, the current sequence is completed as with Replacement selection. When a new sequence is initiated, the records from the reservoir are read first.

It is assumed that the reservoir is on secondary storage, as, if main memory is available, pure Replacement selection with a bigger buffer is always better. If the reservoir is in secondary storage, there is a cost associated with its usage, and there is an interesting trade-off: for a larger reservoir, more records will be passed through it, but longer sequences will result and fewer merging passes will be required.

reservoir size	average run length
$M/2$	$2.15553...M$
M	$2.71828...M$
$3M/2$	$3.16268...M$
$2M$	$3.53487...M$
$5M/2$	$3.86367...M$
$3M$	$4.16220...M$

Table 4.8: Average run lengths for Natural selection

If the number of passes in the sorting algorithm is

$$E[P_n] = \log_b n + O(1)$$

then the optimal reservoir size is the value r which minimizes

$$-\log_b L(r) + \frac{r}{L(r)}$$

where $L(r)$ is the average run length with reservoir size r. Table 4.9 shows some values for the optimal reservoir size. The above function is very "flat" around its minimum, so large variations in the reservoir size do not depart significantly from the optimum.

b	reservoir	average run length	passes saved
2	$6.55M$	$5.81M$	1.409
3	$2.22M$	$3.68M$	0.584
4	$1.29M$	$2.99M$	0.358

Table 4.9: Optimum reservoir sizes for various sorting orders

References:
[Frazer *et al.*, 72], [Knuth, 73], [McKellar *et al.*, 72], [Ting *et al.*, 77].

4.4.0.3 Alternating selection

Some algorithms require that sequences be stored in ascending and descending order alternatively. The Replacement selection algorithm can be used for this purpose with a single change: the last **if** statement should be

```
if (Buff[hbot].k < lastout ↑ direction == 'a' )
```

where *direction* is a global variable which contains the letter "a" or the letter "d". The priority queue functions should also use this global indicator.

The alternation between ascending and descending sequences should be commanded by the function *nextfile.* As a general rule, longer sequences are obtained when the direction is not changed, so the function *nextfile* should be designed to minimize the changes in direction. If the direction is changed for every run, the average length of run is

$$E[n_i] = \frac{3M}{2} + o(1)$$

References:
[Knuth, 63], [Knuth, 73].

4.4.0.4 Merging phase

During the merging phase, the only difference between the algorithms is the selection of the input and output units. The function *merge* merges one run from all the input units (files with the letter "i" in their corresponding FilStat[] entry) into the file given as parameter. This function will be used by all the external merge sorts.

Merge one ordered sequence

```
merge( out )
int out;

{
int i, isml;
typekey lastout;
extern struct rec LastRec[];
extern char FilStat[];

lastout = MinimumKey;
LastRec[0].k = MaximumKey;
while (TRUE) {
        isml = 0;
        for (i=1; i≤maxfiles; i++)
                if ( FilStat[i]=='i' && !Eof(i) &&
                        LastRec[i].k ≥ lastout &&
                        LastRec[i].k < LastRec[isml].k )
                        isml = i;
```

```
        if ( isml==0 )      {
             for (i=1; i<=maxfiles; i++)
                  if ( FilStat[i]== 'i' && !Eof(i)) return(0);
             return( 'done' );
             }
        WriteFile( out, LastRec[isml] );
        lastout = LastRec[isml].k;
        LastRec[isml] = ReadFile(isml);
        }
};
```

Merge uses the global record array *LastRec*. This array contains the last record read from every input file. When all the input files are exhausted simultaneously, this function returns the word *done*.

General References:
[Barnett, 73], [Flores, 69], [Frazer *et al.*, 72], [Friend, 56], [Gotlieb, 63], [Hyafil *et al.*, 74], [Lorin, 75], [Martin, 71], [McCulloch, 82], [Munro *et al.*, 80], [Schlumberger *et al.*, 73], [Tan *et al.*, 82].

4.4.1 Balanced merge sort

Balanced merge sorting is perhaps the simplest scheme for sorting external files. The files are divided into two groups, and every pass merges the runs from one of the groups while distributing the output into the other group of files.

Let T be the number of sequential files available and let P_n^T denote the number of passes necessary to sort n runs or strings. Then we have:

$$P_n^T = \lceil 2 \log_{\lfloor T/2 \rfloor \lceil T/2 \rceil} n \rceil$$

Balanced merge sort

```
sort()
{
int i, runs;
extern int maxfiles, unit;
extern char FilStat[];
extern struct rec LastRec[];

/*** Initialize input/output files ***/
OpenRead( 1 );
for (i=2; i<=maxfiles; i++)
     if ( i <= maxfiles/2 )      FilStat[i] = '-';
          else   OpenWrite( i );
distribute();
```

```
    do {   /*** re-assign files ***/
           for (i=1; i≤maxfiles; i++)
                  if (FilStat[i] == 'o') {
                         OpenRead( i );
                         LastRec[i] = ReadFile(i);
                         }
                  else   OpenWrite( i );
           for (runs=1; merge(nextfile()) ≠ 'done'; runs++);
           } while (runs>1);
    return( unit );
    };
```

The function that performs the selection of the files to output the runs is very simple and just alternates between all possible output files.

Selection of next file for Balanced merge sort

```
nextfile()
{extern int maxfiles, unit;
extern char FilStat[];

do unit = unit%maxfiles + 1;
       while ( FilStat[unit] ≠ 'o' );
return( unit );
};
```

For simplicity, the current output unit number is kept in the global variable *unit*.

For some particular values of n and T, the Balanced merge may not be optimal, for example $P_9^5 = 5$, but an un-balanced merge can do it in 4 passes. Also it is easy to see that $P_n^T = 2$ for $n \le T-1$. The difference between the optimal and normal Balanced merge is not significant.

Table 4.10 shows the maximum number of runs that can be sorted in a given number of passes for the optimal arrangement of Balanced merge sort.

4.4.2 Cascade merge sort

Cascade merge sort distributes the initial sequences or runs among the output files in such a way that during the merging phase the following pattern can be maintained: each *merging pass* consists of the merging of $T-1$ files into one, until one of the input files is exhausted, then the merging of the $T-2$ remaining files into the emptied one, and so on. The final pass finds one

	number of passes				
Files	3	4	5	6	7
3	2	4	4	8	8
4	4	9	16	32	64
5	6	18	36	108	216
6	9	32	81	256	729
7	12	50	144	576	1728
8	16	75	256	1125	4096
10	25	147	625	3456	15625

Table 4.10: Maximum number of runs sorted by Balanced merge sort

sequence in every file and merges them into a single file.

A *perfect distribution* is a set of numbers of runs which allow this process to be carried to the end without ever having two files exhausted at the same time, except when the process is completed. Perfect distributions depend on the number of files, T, and the number of merging steps, k. E.g. $\{0,3,5,6\}$ is a perfect distribution for $T=4$ and $k=3$.

Let $\{0, s_k^1, s_k^2, \ldots, s_k^{T-1}\}$ be a perfect distribution for k merging steps and T files, then

$$s^i(z) = \sum_k s_k^i z^k = \frac{4}{2T-1} \sum_{-T/2<k<\lfloor T/2 \rfloor} \frac{\cos\alpha_k \cos((T-i)\alpha_k)}{1 - \frac{z}{2\sin\alpha_k}}$$

where $\alpha_k = \dfrac{(4k+1)\pi}{4T-2}$

$$s_k^i = \sum_{j=T-i}^{T-1} s_{k-1}^j; \qquad s_k^0 = 0$$

$$s_k^i \approx \frac{4}{2T-1}\cos\frac{\pi}{4T-2} \times \cos\frac{(T-i)\pi}{4T-2} \times \left(2\sin\frac{\pi}{4T-2}\right)^{-k}$$

$$\frac{1}{2\sin\frac{\pi}{4T-2}} = \frac{2T-1}{\pi} + \frac{\pi}{24(2T-1)} + O(T^{-3})$$

Let t_k be the total number of runs sorted by a T-file Cascade merge sort in k merging passes or the size of the kth perfect distribution. Then

$$t(z) = \sum_k t_k z^k = \frac{4}{2T-1} \sum_{-T/2<k<\lfloor T/2 \rfloor} \sin 2T\alpha_k \; \cos\alpha_k \; (2\sin\alpha_k)^{-(k+1)}$$

$$t_k \approx \frac{4}{2T-1}\sin\frac{2T\pi}{4T-2} \; \cos\frac{\pi}{4T-2} \; \left(2\sin\frac{\pi}{4T-2}\right)^{-(k+1)}$$

$$k = \log_{2T/\pi} n\pi/4 \left(1 + \frac{1}{2T\ln 2T/\pi}\right) + O(T^{-2}\log n)$$

Table 4.11 shows the maximum number of runs sorted by Cascade merge sort for various values of T and k.

	number of passes				
Files	3	4	5	6	7
3	3	7	13	23	54
4	6	14	32	97	261
5	10	30	85	257	802
6	15	55	190	677	2447
7	21	91	371	1547	6495
8	28	140	658	3164	15150
10	45	285	1695	10137	62349

Table 4.11: Maximum number of runs sorted by Cascade merge sort

References:
[Knuth, 73], [Kritzinger *et al.*, 74].

4.4.3 Polyphase merge sort

Polyphase merge sort distributes the initial sequences or runs among the output files in such a way that during the merging phase all merges are done from $T-1$ files into 1. Once the proper distribution has been obtained, the merge proceeds from $T-1$ to 1 until one of the input files is exhausted. At this point the output file is rewound and the empty file is opened for output. The merging continues until the whole file is merged into a single sequence.

A *perfect distribution* is a set of numbers of runs which allow this process to be carried to the end without ever having two files exhausted at the same time, except when the process is completed. Perfect distributions depend on the number of files, T, and the number of merging steps, k. Perfect numbers are a generalization of Fibonacci numbers. E.g. $\{0,2,3,4\}$ is a perfect distribution for $T=4$ and $k=3$.

Let $\{0,s_k^1,s_k^2,\ldots,s_k^{T-1}\}$ be a perfect distribution for k merging steps and T files, then

$$s^i(z) = \sum_k s_k^i z^k = \frac{(z^i-1)z}{2z-1-z^T}$$

$$s_k^i = s_{k-1}^{i-1} + s_{k-1}^{T-1}; \qquad s_k^0 = 0$$

$$s_k^i \approx \frac{1-\alpha_T^{-i}}{2-2T+T\alpha_T}(\alpha_T)^k$$

where $1/\alpha_T$ is the smallest positive root of $2z-1-z^T=0$ and

$$\alpha_T = 2 - \frac{2}{2^T-T+1} + O(T^2 8^{-T})$$

Let t_k be the total number of runs sorted by a T-file polyphase merge in k merging steps, or the size of the kth perfect distribution, then

$$t(z) = \sum_k t_k z^k = \frac{(z^T-Tz+T-1)z}{(2z-1-z^T)(z-1)}$$

$$t_k \approx \frac{T-2}{2-2T+T\alpha_T}(\alpha_T)^k$$

The number of merging steps, M_n, for a perfect distribution with n sequences is then

$$M_n = \log_{\alpha_T}\left(\frac{n(2-2T+T\alpha_T)}{T-2}\right) + o(1)$$

$$\approx \left(1+\frac{1}{2^T \ln 2}\right)\log_2 n + 1 - \log_2(T-2) + O(T2^{-T}) + O(n^{-\varepsilon})$$

for some positive ε.

Let r_k be the total number of runs passed (read and written) in a k-step, perfect distribution, merge. Then

$$r(z) = \sum_k r_k z^k = \frac{(z^T - Tz + T - 1)z}{(2z-1-z^T)^2}$$

$$r_k \approx \frac{\alpha_T - 1}{(2-2T+T\alpha_T)^2}\left((T-2)k + \frac{(\alpha_T-2)T(T^2-2T+2)+2T}{2-2T+T\alpha_T}\right)(\alpha_T)^k$$

Let P_n be the total number of passes of the entire file required to sort n initial runs in k merging steps. Then

$$P_n = \frac{\alpha_T - 1}{2-2T+T\alpha_T}k + O(1)$$

$$\approx \frac{1}{2}\left(1+\frac{T-2+1/\ln 2}{2^T}\right)\log_2 n + 2 - \frac{\log_2(T-2)}{2} + \frac{1}{T-2} + O(T^2 2^{-T})$$

When the actual number of sequences or runs is not a perfect distribution, the sequences can be increased by *dummy sequences* (empty sequences) arbitrarily inserted in the files. Since it is possible to predict how many times each sequence will be processed, we can insert the dummy sequences in those positions which are processed the largest number of times. Of course, the sequences are not literally "inserted" in the files, since the files are assumed to be in sequential devices and no merging is done with these. The selection of best placement of dummy sequences together with the selection of the best possible order (any number of merges larger or equal to the minimum required) gives rise to the *Optimal polyphase merge*.

Polyphase merge sort

```
sort()
{
int i, j, some;
extern int maxfiles, maxruns[], actruns[];
extern struct rec LastRec[];
```

```
/*** Initialize input/output files ***/
OpenRead( 1 );
for (i=2; i≤maxfiles; i++)        OpenWrite( i );

/*** Initialize maximum and actual count of runs ***/
for (i=0; i≤maxfiles; i++)        maxruns[i] = actruns[i] = 0;
maxruns[0] = maxruns[maxfiles] = 1;
distribute();

/*** Initialize merging phase ***/
for (i=2; i≤maxfiles; i++)
        { OpenRead( i ); LastRec[i] = ReadFile(i); }
for (i=1; maxruns[0]>1; i = (i%maxfiles)+1 ) {
        OpenWrite( i );
        while ( maxruns[ (i%maxfiles)+1 ] > 0 ) {
                for (j=1; j≤maxfiles; j++)
                        if ( j≠i ) {
                                if ( maxruns[j]>actruns[j] )
                                        FilStat[j] = '-';
                                else {  FilStat[j] = 'i';
                                        actruns[j]--;
                                        some = TRUE;
                                        }
                                maxruns[j]--;      maxruns[0]--;
                                }
                maxruns[i]++;      maxruns[0]++;
                if (some) { merge(i); actruns[i]++; }
                }
        OpenRead( i );      LastRec[i] = ReadFile(i);
        };
return( i==1 ? maxfiles : i-1 );
};
```

Selection of next file for Polyphase merge sort

```
nextfile()
{extern int maxfiles, maxruns[], actruns[];
int i, j, inc;

actruns[0]++;
if (actruns[0]>maxruns[0]) {
        /*** Find next perfect distribution ***/
        inc = maxruns[maxfiles];
```

```
        maxruns[0] += (maxfiles-2) * inc;
        for (i=maxfiles; i>1; i--)
                maxruns[i] = maxruns[i-1] + inc;
        }
j = 2;
/*** select file farthest from perfect ***/
for (i=3; i<=maxfiles; i++)
        if ( maxruns[i]-actruns[i] > maxruns[j]-actruns[j] ) j = i;
actruns[j]++;
return( j );
};
```

Table 4.12 shows the maximum number of runs sorted by Polyphase merge sort for various numbers of files and passes.

	number of passes				
Files	3	4	5	6	7
3	3	7	13	26	54
4	7	17	55	149	355
5	11	40	118	378	1233
6	15	57	209	737	2510
7	19	74	291	1066	4109
8	23	90	355	1400	5446
10	31	122	487	1942	7737

Table 4.12: Maximum number of runs sorted by Polyphase merge sort

References:

[Colin *et al.*, 80], [Er *et al.*, 82], [Gilstad, 60], [Gilstad, 63], [Knuth, 73], [Kritzinger *et al.*, 74], [MacCallum, 73], [Malcolm, 63], [Manker, 63], [McAllester, 64], [Shell, 71], [Zave, 77].

4.4.4 Oscillating merge sort

Oscillating sort interleaves the distribution or dispersion phase together with the merging phase. To do this it is required that the input/output devices be able to:

(a) read backwards;

(b) switch from writing to reading backwards;

(c) switch from reading backwards to writing, without rewinding and without destroying what is at the beginning of the file.

Oscillating sort will always do the mergings reading backwards from $T-2$ units into one. Furthermore, the merging steps are done with balanced files, in the sense that their expected number of records in each is the same. A sequence, ascending (or descending), with $(T-2)n$ initial runs is constructed by a $T-2$-way merge from $T-2$ sequences (each containing n runs) in descending (or ascending) order.

A perfect distribution for Oscillating sort can be produced when $n = (T-2)^k$. The number of passes required to sort n initial runs is:

$$P_n = \lceil \log_{T-1} n \rceil + 1$$

Oscillating sort

```
procedure sort( n, unit, direction : integer );
var     i, r : integer;

begin
if n=0 then   {*** Mark as dummy entry ***}
        FilStat[unit] := '-'
else if n=1 then
        ReadOneRun( unit, direction )

else  for i:=1 to T-2 do begin
        r := n div (T-i-1);
        n := n-r;
        sort( r, (unit+i-2) mod T + 2, -direction );
        MergeOneRunInto( unit, -direction )
        end
end;
```

Table 4.13 shows the maximum number of runs sorted by Oscillating sort or any of its modified versions, for various numbers of files and passes. Note that since the input unit remains open during most of the sorting process, it is not possible to sort with less than 4 units.

	number of passes				
Files	3	4	5	6	7
3	-	-	-	-	-
4	4	8	16	32	64
5	9	27	81	243	729
6	16	64	256	1024	4096
7	25	125	625	3125	15625
8	36	216	1296	7776	46656
10	64	512	4096	32768	262144

Table 4.13: Maximum number of runs sorted by Oscillating merge sort

References:
[Goetz *et al.*, 63], [Knuth, 73], [Lowden, 77], [Sobel, 62].

4.4.5 External Quicksort

External Quicksort is a completely different type of procedure for external sorting. The basic algorithm is the same as the internal Quicksort: the file is split into two parts, the lower and the upper, and the procedure is applied to these recursively. Instead of keeping one single record to do the splitting, this procedure keeps an array of size M of splitting elements. This array of records is maintained dynamically as the splitting phase progresses. Its goal is to produce an even split and to place as many records as possible in their final position (all records in the buffer will be placed in their final location).

For a file consisting of random records, assuming that each pass leaves the records in random order,

$$E[P_n] = \frac{\ln n}{H_{2M} - H_M} + O(1)$$

External Quicksort

```
sort( a, b )
int a, b;

{int i, j, rlow, rupp, wlow, wupp, InBuff;
typekey MaxLower, MinUpper;
struct rec LastRead;
extern struct rec Buff[ ];

while ( b>a ) {
        rupp = wupp = b;
        rlow = wlow = a;
        InBuff = 0;
        MaxLower = MinimumKey;
        MinUpper = MaximumKey;
        i = a-1;
        j = b+1;
        /*** Partition the file ***/
        while ( rupp ≥ rlow ) {
                if ( rlow-wlow < wupp-rupp )
                        LastRead = ReadDirect( rlow++ );
                else    LastRead = ReadDirect( rupp-- );
                if ( InBuff < M ) {
                        Buff[ InBuff++ ] = LastRead;
                        intsort( Buff, 0, InBuff-1 );
                        }
                else {
```

```
                    if ( LastRead.k > Buff[M-1].k ) {
                         if ( LastRead.k > MinUpper ) j = wupp;
                                        else  MinUpper = LastRead.k;
                         WriteDirect( wupp--, LastRead);
                         }
                    else if ( LastRead.k < Buff[0].k ) {
                         if ( LastRead.k < MaxLower ) i = wlow;
                                        else  MaxLower = LastRead.k;
                         WriteDirect( wlow++, LastRead);
                         }
                    else if ( wlow-a < b-wupp ) {
                         WriteDirect( wlow++, Buff[0] );
                         MaxLower = Buff[0].k;
                         Buff[0] = LastRead;
                         intsort( Buff, 0, M-1 );
                         }
                    else { WriteDirect( wupp--, Buff[M-1] ) ;
                         MinUpper = Buff[M-1].k;
                         Buff[M-1] = LastRead;
                         intsort( Buff, 0, M-1 );
                         }
                    }
               }
          while ( InBuff>0 ) WriteDirect( wupp--, Buff[--InBuff] );

          /*** sort the shortest subfile first ***/
          if (i-a < b-j)       { sort(a,i);   a = j; }
               else    { sort(j,b);   b = i; }
          }
     return( 1 );
     };
```

The most noticeable differences between internal and external quicksort are:

(i) the records kept in the buffer are maintained as close to the centre as possible, i.e. deletions are done on the left or on the right depending on how many records were already passed to the left or right.

(ii) the reading of records is also done as balanced as possible with respect to the writing positions. This is done to improve the performance when the file is not random, but slightly out of order.

(iii) two key values are carried during the splitting phase: *MaxLower* and *MinUpper*. These are used to determine the largest interval which can be guaranteed to be in order. By this mechanism it is possible to sort a totally ordered or reversely ordered file in a single pass.

The function *intsort* is any internal sorting function. Its complexity is not crucial as this function is called about $M \ln n$ times per pass of size n. An internal sorting function which does little work when the file is almost totally sorted is preferred (e.g. Linear insertion sort).

Table 4.14 shows simulation results on external Quicksort. From these results we find that the empirical formula

$$E[P_n] \;=\; \log_2(n/M) - 0.924$$

gives an excellent approximation for files with 1000 or more elements.

n	$M=5$	$M=10$	$M=20$
100	3.5272 ± 0.0011	2.73194 ± 0.00076	2.09869 ± 0.00090
500	5.7057 ± 0.0015	4.74526 ± 0.00077	3.88463 ± 0.00057
1000	6.6993 ± 0.0021	5.69297 ± 0.00095	4.77862 ± 0.00059
5000	9.0555 ± 0.0051	7.9773 ± 0.0016	6.99252 ± 0.00063
10000	10.0792 ± 0.0071	8.9793 ± 0.0026	7.97913 ± 0.00090

Table 4.14: Simulation results (number of passes) for external Quicksort

For very large internal buffers, a double-ended priority queue should be used, instead of the function *intsort*.

External Quicksort requires an external device which supports direct access. This sorting procedure sorts records "in-place", that is, no additional files are required. External Quicksort seems to be an ideal sorting routine for direct access files.

This version of Quicksort will have an improved efficiency when sorting partially ordered files.

References:

[Monard, 80].

Chapter 5: SELECTION ALGORITHMS

5.1 Priority queues

We define *priority queues* as recursive data structures where an order relation is established between a node and its descendants. Without loss of generality this order relation will require that the keys in parent nodes be greater than or equal to keys in the descendant nodes. Consequently the root or head of the structure will hold the maximum element.

The algorithms that operate on priority queues need to perform two basic operations: add an element into the queue; extract and delete the maximum element of the queue. Additionally we may require other operations: construct a priority queue from a set of elements; delete and insert a new element in a single operation; inspect (without deleting) the maximum element and merge two queues into a single priority queue. Certainly some of these operations may be built using others. For each algorithm we will describe the most efficient or basic ones.

The typical calling sequence for these functions in Pascal is:

---

procedure *insert*(*new* : *typekey;* **var** *pq* : *queue*);
function *extract*(**var** *pq* : *queue*) : *typekey;*
function *inspect*(*pq* : *queue*) : *typekey;*
procedure *delete*(**var** *pq* : *queue*);
function *merge*(*a, b* : *queue*) : *queue;*
procedure *delinsert*(*new* : *typekey;* **var** *pq* : *queue*);

For the C implementation, the procedures which use **var** parameters are changed into functions which return the modified priority queue.

For some applications we may superimpose priority queue operations with the ability of: searching for any particular element; searching for the successor (or predecessor) of a given element; deleting an arbitrary element, etc.

Searching structures which accept lexicographical ordering may be used as priority queues. E.g. a binary search tree may be used as a priority queue. To add an element we use the normal insertion algorithm; the minimum is in the leftmost node of the tree; the maximum is in the rightmost node.

In all cases C_n^I will denote the number of comparisons required to insert an element into a priority queue of size n, C_n^E the number of comparisons to extract the maximum element and reconstruct the priority queue and C_n^C the number of comparisons needed to construct a priority queue from n elements.

5.1.1 Sorted/unsorted lists

A *sorted list* is one of the simplest priority queues. The maximum element is the head of the list. Insertion is done after a sequential search finds the correct location. This structure may also be constructed using any list-sorting algorithm.

$$C_n^E = 0$$

$$E[C_n^I] = \frac{n(n+3)}{2(n+1)}$$

$$I_n = \frac{n(n+5)}{6}$$

where I_n is the average number of records inspected for all sequences of n operations which start and finish with an empty queue.

Sorted list insertion

```
list insert( new, pq )
list new, pq;

{struct rec r;
list p;
r.next = pq;
p = &r;
while ( p→next ≠ NULL && p→next→k > new→k ) p = p→next;
new→next = p→next;
p→next = new;
return( r.next );
};
```

Sorted list deletion

```
list delete( pq )
list pq;

{if (pq==NULL) Error /* Delete from empty PQ */;
else return( pq→next );
};
```

A sorted list used as a priority queue is inefficient for insertions, it requires $O(n)$ operations. However it may be a good choice in the following situations:

(i) very few elements in the queue;

(ii) special distributions which will produce insertions near the head of the list;

(iii) no insertions at all (all elements are available and sorted before any extraction is done).

An *unsorted list*, at the other extreme, provides very easy addition of elements, but a costly extraction or deletion.

$$C_n^E = n$$

$$C_n^I = 0$$

Unsorted list insertion

```
list insert( new, pq )
list new, pq;

{new→next = pq;
return( new );}
```

Unsorted list deletion

```
list delete( pq )
list pq;

{struct rec r;
list p, max;
if (pq==NULL)      Error /* Deleting from empty PQ */;
else    {r.next = pq;
        max = &r;
        for (p=pq; p→next ≠ NULL; p=p→next)
                if (max→next→k < p→next→k) max = p;
        max→next = max→next→next;
        return( r.next );
        }
};
```

Unsorted list inspection

```
typekey inspect( pq )
list pq;

{list p;
typekey max;
if (pq==NULL)     Error /* Empty Queue */;
else    {max = pq→k;
        for (p=pq→next; p≠NULL; p=p→next)
                if (max < p→k) max = p→k;
        return( max );
        }
};
```

An unsorted list may be a good choice in the following situations:

(i) the elements are already placed in a list by some other criteria;

(ii) there are very few deletions.

Merging sorted lists is an $O(n)$ process; merging unsorted lists is also an $O(n)$ process unless we have direct access to the tail of one of the lists.

References:

[Nevalainen *et al.*, 79].

5.1.2 P-trees

P-trees or priority-trees are binary trees with a particular ordering constraint which makes them suitable for priority queue implementations. This ordering can be best understood if we tilt the binary tree 45° clockwise and let the left pointers become horizontal pointers and the right pointers become vertical. For such a rotated tree the ordering is lexicographical.

We also impose the condition that the maximum and minimum elements of the tree both be on the leftmost branch, and so on recursively. This implies that any leftmost node does not have right descendants.

The top of the queue, the maximum in our examples, is kept at the leftmost node of the tree. The minimum is kept at the root. This requires some additional searching to retrieve the top of the queue. If we keep additional pointers and introduce pointers to the parent node in each node, the deletion and retrieval of the top element become direct operations. In any case, a deletion does not require any comparisons, only pointer manipulations.

Let L_n be the length of the left path in a queue with n elements. For each node inspected a key-comparison is done. Then for a queue built from n random keys:

$$E[L_n] = 2H_n - 1$$

$$E[C_n^I] = \frac{H_{n+1}^2}{3} + \frac{10H_{n+1}}{9} - \frac{H_{n+1}^{(2)}}{3} - \frac{28}{27} \qquad (n \geq 2)$$

where $H_n = \sum_{i=1}^{n} 1/i$ denotes harmonic numbers and $H_n^{(2)} = \sum_{i=1}^{n} 1/i^2$ denotes biharmonic numbers.

P-tree insertion

```
tree insert( new, pq )
tree new, pq;

{
tree p;
if ( pq == NULL ) return( new );
else if ( pq→k ≥ new→k ) {
                /*** Insert above subtree ***/
                new→left = pq;
                return( new );
                }
else    {
        p = pq;
        while ( p→left ≠ NULL )
                if ( p→left→k ≥ new→k ) {
                        /*** Insert in right subtree ***/
                        p→right = insert( new, p→right );
                        return( pq );
                        }
                else    p = p→left;
        /*** Insert at bottom left ***/
        p→left = new;
        };
return( pq );
};
```

P-tree deletion of maximum

```
tree delete( pq )
tree pq;
{
if ( pq == NULL ) Error /*** deletion on an empty queue ***/;
else if ( pq→left == NULL )     return( NULL );
else if ( pq→left→left == NULL ) {
```

```
        pq→left = pq→right;
        pq→right = NULL;
        }
else    pq→left = delete( pq→left );
return( pq );
};
```

P-tree, retrieval of head of queue

```
typekey inspect( pq )
tree pq;
{
if ( pq == NULL ) Error /*** Inspecting an empty queue ***/;
while ( pq→left ≠ NULL ) pq = pq→left;
return( pq→k );
};
```

With a relatively small change, P-trees allow the efficient extraction of the minimum as well as the maximum, so this structure is suitable for handling *double-ended* priority queues.

This priority queue is stable; equal keys will be retrieved first-in first-out.

Table 5.1 contains exact results (rounded to 6 digits) for the length of the leftmost path and for the number of node inspections during insertions for P-trees. Simulation results are in excellent agreement with the theoretical ones.

n	$E[C_n^C]$	$E[L_n]$
5	7.66667	3.56667
10	27.1935	4.85794
50	347.372	7.99841
100	939.017	9.37476
500	8207.70	12.5856
1000	20001.3	13.9709
5000	147948.6	17.1890
10000	342569.2	18.5752

Table 5.1: Exact results for P-trees

References:
[Jonassen *et al.*, 75], [Nevalainen *et al.*, 78].

5.1.3 Heaps

A *heap* is a perfect binary tree represented implicitly in an array. This binary tree has the priority queue ordering: the key in the parent node is greater than or equal to any descendant key. The tree is represented in an array without the use of pointers. The root is located in position 1. The direct descendants of the node located in position i are those located in $2i$ and $2i+1$. The parent of node i is located at $\lfloor i/2 \rfloor$. The tree is "perfect" in the sense that a tree with n nodes fits into locations $1...n$. This forces a breadth-first, left-to-right filling of the binary tree.

For the Williams' insertion algorithm, let C_n^I denote the number of comparisons and M_n the number of interchanges needed to insert the $n+1^{st}$ element, then

$$1 \le C_n^I \le \lfloor \log_2 n \rfloor \qquad (n>1)$$

$$E[M_n] = E[C_n^I] - \frac{n-1}{n}$$

For an insertion into a random heap (all possible heaps being equally likely), when n is in the range $2^{k-1}-1 \le n < 2^k-1$ we have:

$$E[C_{2^k-2}^I] \le E[C_n^I] \le E[C_{2^{k-1}-1}^I]$$

$$E[C_{2n}^I] < E[C_{2n-1}^I] \qquad (n>1)$$

$$E[C_{2^k-2}^I] = 2 + O(k2^{-k})$$

$$E[C_{2^{k-1}-1}^I] = 2.60669... + O(k2^{-k})$$

A heap built by random insertions using Williams' insertion algorithm is not a random heap.

Williams' heap-insertion algorithm

```
procedure insert( new : ArrayEntry; var r : RecordArray );
var     i, j : integer;
        flag : boolean;

begin
n := n+1;
j := n;
flag := true;
while flag and (j>1) do begin
        i := j div 2;
        if r[i].k ≥ new.k then flag := false
                else begin r[j] := r[i]; j := i end
        end;
r[j] := new
end;
```

If all the elements are available at the same time, we can construct a heap more efficiently using Floyd's method. In this case

$$n-1 \le C_n^C \le 2n-2\nu(n)-\phi(n)$$

$$E[C_{2^k-1}^C] = (\alpha_1+2\alpha_2-2)2^k - 2k - 1 - \frac{6k+5}{9\ 2^k} + O(k4^{-k})$$

$$E[C_n^C] = 1.88137...n + O(\log n)$$

where $\alpha_1 = \sum_{k\ge1} \frac{1}{2^k-1} = 1.60669...$ and $\alpha_2 = \sum_{k\ge1} \frac{1}{(2^k-1)^2} = 1.13733...$

$$0 \le M_n^C \le n - \nu(n)$$

$$E[M_{2^k-1}^C] = (\alpha_1+\alpha_2-2)2^k - k - \frac{3k+4}{9\ 2^k} + O(k4^{-k})$$

$$E[M_n^C] = 0.74403...n + O(\log n)$$

where $\nu(n)$ is the number of 1's in the binary representation of n and $\phi(n)$ is the number of trailing 0's of the binary representation of n.

Floyd's heap-construction algorithm

```
procedure siftup( var r : RecordArray; i,n : integer );

var     j : integer;
        tempr : ArrayEntry;
begin
        while 2*i<=n do begin
                j := 2*i;
                if j<n then
                        if r[j].k < r[j+1].k then j := j+1;
                if r[i].k < r[j].k then begin
                        tempr := r[j];
                        r[j] := r[i];
                        r[i] := tempr;
                        i := j
                        end
                else    i := n+1
                end
end;
for i := (n div 2) downto 1 do siftup(r,i,n);
```

Worst case lower and upper bounds:

$$C_n^I = \lfloor \log_2(\lfloor \log_2 n \rfloor + 1) \rfloor + 1$$

$$C_n^E = \lfloor \log_2 n \rfloor + g(n) + O(1)$$

where $g(0)=0$ and $g(n)=1+g(\lfloor \log_2 n \rfloor)$.

$$C_n^C = \frac{13}{8} n + O(\log n)$$

Average lower bound:

$$1.36443...n + O(\log n) \le C_n^C$$

Extraction and reorganization:

$$C_n^E \le 2\lfloor \log_2(n-1) \rfloor - \rho(n-1)$$

where $\rho(n)$ is 1 if n is a power of 2, 0 otherwise.

Heap extraction and reorganization

```
function extract( var r : RecordArray ) : typekey;
begin
if n<1 then Error {*** extracting from an empty Heap ***}
else    begin
        extract := r[1].k;
        r[1] := r[n];
        n := n-1;
        siftup( r, 1, n )
        end
end;
```

For a random Delete-Insert operation into a random heap we have:

$$2 \le C_n^E \le 2\lfloor \log_2 n \rfloor - \rho(n)$$

$$E[C_n^E] = \frac{2\big((n+1)k - \lfloor n/2 \rfloor - 2^k\big)}{n}$$

where $k = \lfloor \log_2 n \rfloor + 1$.

$$E[C_{2^k-1}^E] = \frac{(2k-3)2^k + 2}{2^k - 1}$$

Heap Delete-Insert algorithm

```
procedure delinsert( new : RecordEntry; var r : RecordArray );
begin
r[1] := new;
```

```
siftup( r, 1, n )
end;
```

The heap does not require any extra storage besides the elements themselves. These queues can be implemented just by using arrays and there are no requirements for recursion.

The insertion and extraction operations are guaranteed to be $O(\log n)$.

Whenever we can allocate vectors to store the records, the heap seems to be an ideal priority queue.

Merging two disjoint heaps is an $O(n)$ operation.

We can generalize the heap to any branch factor b other than two; in this case the parent of node i is located at $\lfloor (i-2)/b \rfloor + 1$ and the descendants are located at $\lceil b(i-1)+2 \rceil$, ..., $\lceil bi+1 \rceil$. This provides a tradeoff between insertion and extraction times: the larger b, the shorter the insertion time and longer the extraction time.

Table 5.2 gives figures for the number comparisons, C_n^C, required to build a heap by repetitive insertions, the number of comparisons required to insert the $n+1^{st}$ element, C_n^I and the number of comparisons required to extract all the elements from a heap constructed in this manner, C_n^D.

n	$E[C_n^C]$	$E[C_n^I]$	$E[C_n^D]$
5	5.133333	1.583333	5.79970 ± 0.00078
10	13.95278	1.667027	25.5400 ± 0.0038
50	96.60725	1.983653	330.165 ± 0.029
100	206.0169	2.135882	850.722 ± 0.062
500	1103.952	2.116126	6501.21 ± 0.26
1000	2237.56 ± 0.59	2.2538 ± 0.0068	14989.06 ± 0.53
5000	11348.8 ± 3.2	2.330 ± 0.015	98310.6 ± 3.2
10000	22749.8 ± 6.6	2.401 ± 0.022	216592.0 ± 6.2

Table 5.2: Complexity of heaps created by insertions

References:
[Doberkat, 81], [Doberkat, 82], [Doberkat, 82], [Floyd, 64], [Gonnet *et al.*, 82], [Gonnet, 76], [Kahaner, 80], [Knuth, 73], [Porter *et al.*, 75], [Williams, 64].

5.1.4 VanEmde-Boas priority queues

VanEmde-Boas priority queues are queues which perform the operations insert, delete, extract maximum or minimum and find predecessor or successor in $O(\log \log N)$ operations. For these queues, N represents the size of the universe of keys and n the actual size of the subset of keys we include in the queue. It makes sense to use these queues when the keys are subsets of the integers $1...N$.

These queues are represented by one of various possible data structures. A queue is either

(a) empty, in which case it is represented by **nil;**

(b) a single element, in which case it is represented by the integer element itself;

(c) over a small universe ($N \le m$), in which case it is represented as a boolean array of size N, or

(d) as a structure composed of a queue of queues. The queue of queues is called the *top* part, and the element queues, which are arranged as an array, are called the *bottom* queues. Additionally we keep the maximum and minimum value occurring in the queue. The sizes of the top and bottom queues are as close to the square root of the cardinality of the universe as possible.

As a hyperrule, these priority queues have the definition:

$$\mathbf{vEB-N : [int, int, vEB-s(N), \{vEB-s(N)\}_1^{s(N)} \; ; \; \{bool\}_1^N \; ; \; int \; ; \; nil}$$

where $\mathbf{s(N)} = \lceil \sqrt{\mathbf{N}} \rceil$. The top queue is a queue on the indices of the bottom array. The index of every non-empty queue in the bottom is a key in the top queue.

VanEmde-Boas priority queue insertion

```
insert( new : integer; var pq );
case pq is nil:
        pq := NewSingleNode( new );
case pq is boolean array:
        turn on corresponding entry;
case pq is single element:
        expand entry to full node;
        seep into next case;
case pq is full node:
        compute index based on "new"
        if bottom[index] ≠ nil then
                        insert in bottom[index]
        else    bottom[index] := NewSingleNode( new );
                insert index in top queue;
        adjust max and min if necessary;
end;
```

VanEmde-Boas priority queue extraction

```
extract( var pq ) : integer;
case pq is nil:
        Error;
case pq is boolean array:
```

```
        Find last true entry;
        if only one entry remains then transform to SingleEntry;
    case pq is single element:
        return element;
        pq := nil;
    case pq is full node:
        return maximum;
        if bottom queue corresponding to maximum is single element
                then    extract from top queue;
                        max := max of bottom[ max of top ];
        else    extract from bottom;
                max := max of bottom;
    end;
```

Let S_n^N be the storage utilized by a queue with n elements from a universe of size N. Then

$$S_n^N = O(min(\sqrt{N} + nN^{1/4}, N))$$

The functions *extract minimum, test membership, find successor* and *find predecessor* can also be implemented in the same time and space.

References:
[vanEmde-Boas *et al.*, 77], [vanEmde-Boas, 75], [vanEmde-Boas, 77].

5.1.5 Pagodas

The pagoda is an implementation of a priority queue in a binary tree. The binary tree is constrained to have priority queue ordering, (parent larger than descendants). The structure of the pointers in the pagoda is peculiar; we have the following organization:

(i) the root pointers point to the leftmost and to the rightmost nodes;

(ii) the right link of a right descendant points to its parent and its left link to its leftmost descendant;

(iii) the left link of a left descendant points to its parent and its right link to its rightmost descendant.

The basic operation in a pagoda is merging two disjoint pagodas, which can be done very efficiently. An insertion is achieved by merging a single element with the main structure; an extraction is done by merging the two descendants of the root.

Merging pagodas is done bottom-up, merging the leftmost path of one with the rightmost path of the other.

Let C_{mn}^M be the number of comparisons needed to merge two pagodas of sizes m and n respectively. Then for pagodas built from random input we have

$$1 \le C_{mn}^M \le m+n-1$$

$$E[C^M_{mn}] = 2(H_n + H_m - H_{m+n})$$

$$1 \leq C^I_n \leq n$$

$$E[C^I_n] = 2 - \frac{2}{n+1}$$

$$0 \leq C^E_n \leq n-2$$

$$E[C^E_n] = 2(H_n - 2 + \frac{1}{n})$$

$$n-1 \leq C^C_n \leq 2n-3$$

$$E[C^C_n] = 2n - 2H_n$$

Merging two pagodas

```
function merge( a, b : tree ) : tree;
var     bota, botb, r, temp : tree;

begin
if a=nil then merge := b
else if b=nil then merge := a
else    begin
        {*** Find bottom of a's rightmost edge ***}
        bota := a↑.right;    a↑.right := nil;
        {*** bottom of b's leftmost edge ***}
        botb := b↑.left;     b↑.left := nil;
        r := nil;
        {*** Merging loop ***}
        while (bota≠nil) and (botb≠nil) do
                if bota↑.k < botb↑.k then begin
                        temp := bota↑.right;
                        if r=nil then bota↑.right := bota
                                else    begin
                                        bota↑.right := r↑.right;
                                        r↑.right := bota
                                        end;
                        r := bota;
                        bota := temp
                        end
                else    begin
                        temp := botb↑.left;
                        if r=nil then botb↑.left := botb
                                else    begin
                                        botb↑.left := r↑.left;
                                        r↑.left := botb
                                        end;
```

```
                    r := botb;
                    botb := temp
                    end;
          {*** one edge is exhausted, finish merge ***}
          if botb=nil then begin
               a↑.right := r↑.right;
               r↑.right := bota;
               merge := a
               end
          else begin
               b↑.left := r↑.left;
               r↑.left := botb;
               merge := b
               end
          end
end;
```

Insertion in a pagoda

```
procedure insert( new : tree; var pq : tree );
begin
new↑.left := new;   new↑.right := new;
pq := merge( pq, new )
end;
```

Deletion of head in a pagoda

```
procedure delete( var pq : tree );
var    le, ri : tree;

begin
if pq=nil then Error {*** Deletion on empty queue ***}
else  begin
      {*** Find left descendant of root ***}
      if pq↑.left = pq then le := nil
            else begin
                 le := pq↑.left;
                 while le↑.left ≠ pq do le := le↑.left;
                 le↑.left := pq↑.left
                 end;
```

```
        {*** Find right descendant of root ***}
        if pq↑.right = pq then ri := nil
             else begin
                  ri := pq↑.right;
                  while ri↑.right ≠ pq do ri := ri↑.right;
                  ri↑.right := pq↑.right
                  end;
        {*** merge descendants ***}
        pq := merge( le, ri )
        end
end;
```

Pagodas are remarkably efficient in their average behaviour with respect to the number of comparisons.

References:
[Francon *et al.*, 78], [Nix, 79].

5.1.6 Binary trees used as priority queues

5.1.6.1 Leftist trees

A leftist tree is a binary tree with a priority queue ordering, which uses a count field at every node. This count field indicates the height (or distance) to the closest leaf. Leftist trees are arranged so that the sub-tree with the shortest path to a leaf is on the right descendant.

These trees are called *leftist* as their left branches are usually taller than their right ones.

An insertion can be done in the path to any leaf, so it is best to do it towards the rightmost leaf which is the closest to the root. A deletion is done through merging the two immediate descendants for the root. Leftist trees allow efficient, $O(\log n)$, merging of different trees.

Leftist tree insertion

```
procedure insert ( new : tree; var pq : tree );

begin
if pq = nil then     pq := new
else if pq↑.k > new↑.k then begin
               insert( new, pq↑.right );
               fixdist( pq )
               end
else   begin
       new↑.left := pq;
       pq := new
```

```
        end
end;
```

Leftist tree deletion

```
function merge( a, b : tree ) : tree;
begin
if a = nil then merge := b
else if b = nil then merge := a
else if a↑.k > b↑.k then begin
        a↑.right := merge( a↑.right, b );
        fixdist( a );
        merge := a
        end
else    begin
        b↑.right := merge( a, b↑.right );
        fixdist( b );
        merge := b
        end
end;

procedure delete ( var pq : tree );
begin
if pq = nil then Error {*** delete on an empty queue ***}
else    pq := merge( pq↑.left, pq↑.right )
end;
```

The function *fixdist* recomputes the distance to the closest leaf by inspecting at the right descendant, if any.

All operations on the leftist trees require $O(\log n)$ time even in the worst case.

Table 5.3 summarizes simulation results on leftist trees. C_n^C indicates the number of comparisons required to build a leftist tree, *dist* indicates the distance from the root to the closest leaf and C_n^D the number of comparisons required to extract all the elements from the tree.

References:

[Crane, 72], [Sleator *et al.*, 83].

n	$E[C_n^C]$	$E[dist]$	$E[C_n^D]$
10	14.5955 ± 0.0099	2.4314 ± 0.0029	11.992 ± 0.010
50	131.44 ± 0.14	3.6807 ± 0.0097	176.056 ± 0.081
100	317.11 ± 0.41	4.211 ± 0.015	469.35 ± 0.18
500	2233.6 ± 4.7	5.497 ± 0.041	3779.2 ± 1.0
1000	5036 ± 14	6.071 ± 0.063	8817.3 ± 2.2
5000	31845 ± 155	7.45 ± 0.16	58797 ± 13
10000	69500 ± 395	7.97 ± 0.23	130312 ± 22

Table 5.3: Simulation results for leftist trees

5.1.6.2 Binary priority queues

We can construct a binary tree with a priority queue ordering instead of a lexicographical ordering. By doing this, most of the algorithms for binary trees can also be used for priority queues. There is a contradiction of goals however. While the best binary tree for searching is a tree as height balanced as possible, the best tree for a priority queue is one which is as thin or as tall as possible. With this in mind we can devise an algorithm to produce rather tall trees.

For simplicity of the algorithms we will impose the following conditions:

(i) the key in the node is larger than any other key in the descendant sub-trees;

(ii) if a sub-tree is non-null, then the left sub-tree is non-null;

(iii) the key in the direct left descendant (if any) is larger than the key in the direct right descendant.

Binary priority queue insertion

```
procedure insert ( new : tree; var pq : tree );

begin
if pq = nil then        pq := new
else if pq↑.k ≤ new↑.k then begin
                 new↑.left := pq;
                 pq := new
                 end
else if pq↑.left = nil then
                 pq↑.left := new
else if pq↑.left↑.k ≤ new↑.k then
                 insert( new, pq↑.left )
else             insert( new, pq↑.right )
end;
```

Binary priority queue deletion

```
procedure delete ( var pq : tree );
var     temp : tree;
begin
if pq = nil then Error {*** deletion on an empty queue ***}
else if pq↑.right = nil then
        pq := pq↑.left
else    begin
        {*** promote left descendant up ***}
        pq↑.k := pq↑.left↑.k;
        delete( pq↑.left );
        {*** rearrange according to constraints ***}
        if pq↑.left = nil then begin
                pq↑.left := pq↑.right;      pq↑.right := nil  end;
        if pq↑.right ≠ nil then
                if pq↑.left↑.k < pq↑.right↑.k then begin
                        {*** descendants in wrong order ***}
                        temp := pq↑.right;
                        pq↑.right := pq↑.left;
                        pq↑.left := temp
                        end
        end
end;
```

Table 5.4 summarizes the simulation results for Binary priority queues. I_n indicates the number of iterations performed by the insertion procedure, C_n^C the number of comparisons to construct the queue and C_n^D the number of comparisons to extract all the elements from the queue.

n	$E[I_n]$	$E[C_n^C]$	$E[C_n^D]$
10	18.3524±0.0079	23.384±0.016	7.1906±0.0071
50	148.56±0.13	232.65±0.26	139.80±0.12
100	353.51±0.40	578.59±0.80	396.03±0.40
500	2463.6±4.9	4287.1±9.3	3472.2±6.0
1000	5536±14	9793±28	8276±18
5000	34827±161	63258±322	56995±204
10000	75931±407	139071±814	127878±569

Table 5.4: Simulation results for Binary tree priority queues

5.1.6.3 Binary search trees as priority queues

Binary search trees, in any of their variations, can be used as priority queues. The maximum is located at the rightmost node and the minimum is located at the leftmost node. The insertion algorithm is almost the same as for Binary search trees, except that we are not concerned about duplicated keys. An extraction is done by deleting the rightmost node which is one of the easy cases of deletion.

The complexity measures for random insertions are the same as those for Binary search trees (section 3.4.1).

Binary search tree insertion

```
procedure insert( new : tree; var t : tree );
begin
if t = nil then        t := new
else if t↑.k < new↑.k then insert( new, t↑.right )
                       else        insert( new, t↑.left )
end;
```

Binary search tree, extraction of maximum

```
function extract( var pq : tree ) : typekey;
begin
if pq=nil then Error {*** Extraction from empty queue ***}
else if pq↑.right = nil then begin
        extract := pq↑.k;
        pq := pq↑.left
        end
else    extract := extract( pq↑.right )
end;
```

Binary search trees used as queues behave as double-ended priority queues, since we can extract both the maximum and the minimum element. Binary search trees are not easy to merge as they require linear time in their total size.

This priority queue is stable; equal keys will be retrieved first-in first-out.

When used for insertions intermixed with extractions, this type of queue tends to degenerate into a skewed tree. For this reason it appears to be much safer to use any type of balanced binary tree.

5.1.7 Binomial queues

Binomial queues use binary decomposition to represent sets of values with special structures of sizes 1, 2, 4, ... 2^k. A structure of size 2^k is called a B_k tree and has the following properties:

(a) the maximum element of the set is the root;

(b) the root has k descendants; one B_0, one B_1, ... one B_{k-1} tree.

B_k trees are the natural structure that arises from a tournament between 2^k players.

Two B_k trees can be joined into a single B_{k+1} tree with one single comparison. Consequently a B_k tree can be constructed using $2^k - 1$ comparisons. This construction is optimal.

A binomial queue of size n is represented as a *forest* of B_k trees where there is at most one B_k tree for each k. This corresponds to the binary decomposition of n. E.g. $n=13=1101_2$ is represented by $\{B_3,B_2,B_0\}$.

The maximum element of a binomial queue can be found by inspecting the head of all of its B_k trees which requires $\nu(n)-1 \le \lfloor \log_2 n \rfloor$ comparisons (where $\nu(n)$ is the number of "1" digits in the binary representation of n).

Two binomial queues can be merged into a single queue by joining all equal-size B_k trees in a process which is identical to binary addition. Merging two queues with sizes m and n requires

$$C_{mn}^{M} = \nu(n) + \nu(m) - \nu(m+n)$$

An insertion of a single element into a queue with n elements is treated as a merge and hence requires

$$C_n^I = \nu(n) + 1 - \nu(n+1)$$

Constructing a binomial queue by repetitive insertions requires

$$C_n^C = n - \nu(n)$$

A deletion of an extraction is accomplished by removing the largest root of the B_k trees and merging all its descendants with the original queue. This operation requires

$$\nu(n) - 1 \le C_n^D \le 2\nu(n) + \lfloor \log_2 n \rfloor - \nu(n-1) - 1$$

Binomial queues can be implemented using binary trees. These implementations are simplified if we include the size of each B_k tree in the root node.

Binomial queues give excellent worst-case behaviour for insertions, constructions by insertions, deletions and merging of queues at the cost of two pointers per entry.

5.1.8 Summary

The following table shows an example of real relative total times for constructing a priority queue with 10007 elements by repetitive insertions and then extracting all its elements.

General References:

[Ajtai *et al.*, 83], [Brown *et al.*, 79], [Flajolet *et al.*, 79], [Flajolet *et al.*, 80], [Itai *et al.*, 80], [Johnson, 75], [Pohl, 75], [Standish, 80].

Algorithm	C	Pascal
Sorted lists	67.3	52.2
Unsorted lists	204.1	85.5
P-trees	6.1	4.8
Heaps	1.0	1.0
Pagodas	4.0	3.0
Leftist trees	7.3	6.8
Binary priority queues	4.6	3.5
B.S.T as priority queues		3.4

Table 5.5: Relative total times for priority queue algorithms

5.2 Selection of kth

The selection of the kth element is defined as searching for an element X in an unordered set such that $k-1$ elements from the set are less than or equal to X and the rest are greater than or equal to X.

Finding the first or last (minimum or maximum) is the most important special case and was treated in the first section of this chapter. Finding the median (or closest to the median) is another special case of selection.

Let $C_{k,n}$ denote the number of comparisons needed to find the kth element in a set containing n unordered elements. Let $C_{k,n}^{MM}$ denote the minimum maximum or minimum worst case number of comparisons for the same problem. For any algorithm we have the lower bound on the average median selection:

$$E[C_{n/2,n}] \geq \frac{3n}{2} + O(1)$$

For the Floyd and Rivest algorithm we have:

$$E[C_{k,n}] = n + min(k,n-k) + O(\sqrt{n})$$

For small k,

$$E[C_{k,n}] \leq n + O(k \ln \ln n)$$

Table 5.6 summarizes the worst-case upper and lower bounds on the problem.

In the algorithms presented below, we assume that all the records are stored in an array. This array can be shuffled if necessary.

5.2.1 Selection by sorting

One of the simplest strategies for selection is to sort all the array and then directly select the desired element.

$C_{k,n}^{MM}$	
	Lower bounds
$k=1$	$n-1$
$k=2$	$n-2+\lceil \log_2 n \rceil$
for any j	
$k=3,\ n=2^j+1$	$n-3+2\lceil \log_2(n-1)\rceil$
$k=3,\ 3\times 2^j<n\leq 4\times 2^j$	$n-4+2\lceil \log_2(n-1)\rceil$
$k=3,\ 2\times 2^j+1<n\leq 3\times 2^j$	$n-5+2\lceil \log_2(n-1)\rceil$
$2k-1\leq n<3k$	$\lfloor \frac{3n+k-5}{2} \rfloor$
$3k\leq n$	$n+k-3+\sum_{j=0}^{k-2} \log_2 \lceil \frac{n-k+2}{k+j} \rceil$
$2k=n$	$\frac{79n}{43}+O(1)$
	Upper bounds
$k=1$	$n-1$
$k=2$	$n-2+\lceil \log_2 n \rceil$
$k\geq 1$	$n-k+(k-1)\lceil \log_2(n-k+2)\rceil$
$2^s(2^{\lceil \log_2 k \rceil}+j)<n-k+2$ and $n-k+2\leq 2^s(2^{\lceil \log_2 k \rceil}+j+1)$ and $\lfloor k/2 \rfloor > j\lceil \log_2 k \rceil$	$n-k+(k-1)\lceil \log_2(n-k+2)\rceil - \lfloor (k-1)/2 \rfloor + j\lceil \log_2 k \rceil$
$2k=n+1$	$3n+O((n\log n)^{3/4})$
$5k\leq n$	$\approx n(1+2^{1-\lceil \log_2(n/5k)\rceil})+5k\lceil \log_2(n/5k)\rceil$

Table 5.6: Upper and lower bounds for k^{th} selection

Selection by sorting

```
function select( i : integer; var r : RecordArray;
                 lo, up : integer ) : typekey;
begin
if (i<lo) or (i>up) then Error {*** Selection out of bounds ***}
else begin
     sort( r, lo, up );
     select := r[i].k
     end
end;
```

This method is expensive for selecting a single element but should be preferred whenever several successive selections are performed.

5.2.2 Selection by tail recursion

This function uses a tail recursion technique. Each iteration starts by selecting a splitter element from the file. The file is then split into two subfiles, those elements with keys smaller than the selected one and those elements with larger keys. In this way, the splitting element is placed in its proper final location between the two resulting subfiles. This procedure is repeated recursively on the subfile which contains the element to be selected.

For a randomly ordered file, the first selection of the kth element will require

$$E[C_{k,n}] = 2\big((n+1)H_n - (n+3-k)H_{n+1-k} - (m+2)H_k + n + 3\big)$$

$$E[C_{n/2,n}] \approx 3.38629...n$$

Selection by tail recursion

```
function select( s : integer; var r : RecordArray;
                 lo, up : integer ) : typekey;

var     i, j : integer;
        tempr : ArrayEntry;
begin
if (s<lo) or (s>up) then Error {*** Selection out of bounds ***}
else begin
        while (up≥s) and (s≥lo) do begin
                i := lo;
                j := up;
                tempr := r[s];  r[s] := r[lo];  r[lo] := tempr;
                {*** Split file in two ***}
                while i<j do begin
                        while r[j].k > tempr.k do
                                j := j-1;
                        r[i] := r[j];
                        while (i<j) and (r[i].k≤tempr.k) do
                                i := i+1;
                        r[j] := r[i]
                        end;
                r[i] := tempr;
                {*** Select subfile ***}
                if s<i then up := i-1
                        else lo := i+1
                end;
        select := r[s].k
        end
end;
```

The above algorithm uses as a splitting element the one located at the selected position. For a random file, any location would provide an equivalently good splitter. However, if the procedure is applied more than once, any other element (e.g. the first) may produce an almost worst-case behaviour.

As selections are done, the array is sorted into order. It is expected that later selections will cost less, although these will always use $O(n)$ comparisons.

Strategies which select, in place, a smaller sample to improve the splittings, cause an almost worst-case situation and should be avoided. Sampling, if done, should not alter the order of elements in the array.

Any of the distributive methods of sorting, e.g. Bucket sort, top-down Radix sort, can be modified to do selection. In all cases the strategy is the same: the sorting algorithms split the file into several sub-files and are applied recursively onto each sub-file (divide and conquer). For selection, we do the same first step, but then we select only the sub-file that will contain the desired element (by counting the sizes of the sub-files) and apply recursion only on one sub-file (tail recursion).

5.2.3 Selection of the mode

The *mode of a set* is defined as the key value which occurs most frequently in a set. The number of times a key value is repeated is called the *multiplicity*. The mode is the key with largest multiplicity.

The selection of the mode is almost trivial when the set is ordered, as a single pass through the set is enough to determine it. The complexity of determining the mode is lower than the complexity of sorting; in fact if n_m is the multiplicity of the mode, then we have the upper and lower bound

$$C_n^{Mode} \approx n \log_2(n/n_m) - n \log_2(\log_2 n - P)$$

where $P = \sum_i n_i/n \log_2 n_i$ and n_i is the multiplicity of the ith different key value.

The following algorithm uses a divide and conquer technique to find the mode.

Determining the mode

```
function mode( S : SetOfKeys ) : SetOfKeys;
var    A, A1, A2, A3 : SetOfKeys;
       Homog, Heter : set of SetOfKeys;
begin
Homog := [];
Heter := [S];
```

```
while LargestCardinality(Heter) > LargestCardinality(Homog) do
      begin
      A := LargestSet( Heter );
      med := median( A );
      split A into A1, A2, A3
              { with elements <med; =med; >med }
      Heter := ( Heter - A ) + A1 + A3;
      Homog := Homog + A2
      end;
LargestSet( Homog );
end;
```

This algorithm requires

$$C_n^{Mode} \approx kn \log_2(n/n_m)$$

comparisons, where kn is the number of comparisons required to find the median among n elements. E.g. using the Schonhage, Paterson and Pippenger median algorithm, $k=3$ and

$$C_n^{Mode} \approx 3n \log_2(n/n_m)$$

in the worst case.

Dobkin and Munro describe an optimal, within lower order terms, algorithm to find the mode. Their optimal algorithm is rather complicated, and mostly of theoretical interest.

References:
[Dobkin *et al.*, 80].

General References:
[Aho *et al.*, 74], [Allison *et al.*, 80], [Blum *et al.*, 73], [Cunto, to app.], [Dobkin *et al.*, 81], [Eberlein, 79], [Floyd *et al.*, 75], [Frederickson *et al.*, 80], [Fussenegger *et al.*, 76], [Fussenegger *et al.*, 79], [Hoare, 61], [Hyafil, 76], [Johnson *et al.*, 78], [Kirkpatrick, 81], [Knuth, 73], [Motoki, 82], [Munro *et al.*, 80], [Munro *et al.*, 82], [Noshita, 74], [Nozaki, 73], [Pratt *et al.*, 73], [Reiser, 78], [Schonhage *et al.*, 76], [Wirth, 76], [Yao *et al.*, 82], [Yap, 76].

Chapter 6: ARITHMETIC ALGORITHMS

6.1 Basic operations, multiplication/division

In this section we will discuss arithmetic algorithms to perform the basic operations. Given that addition and subtraction are relatively straightforward, we will concentrate on multiplication/division and other operations.

Our model of computation can be called *multiple-precision,* as we are interested in describing arithmetic operations in terms of operations in a much smaller domain. For example, some algorithms may implement decimal operations using ascii characters as basic symbols, or we may implement extended precision using basic integer arithmetic, or integer arithmetic using bits, etc. Without loss of generality we will call the basic unit of implementation *digit,* and a logical collection of digits *number.* Our complexity measures will be given in number of operations on digits as a function of the number of digits involved.

Let $M(n)$ denote the complexity of multiplying two n-digit numbers and let $Q_{f(x)}(n)$ denote the complexity of computing the function $f(x)$ with n-digit precision. ($Q_{\times}(n) = M(n)$). Then we have the following results:

$$Q_{+}(n) = Q_{-}(n) = Q_{kx}(n) = O(n)$$

for an integer constant k. The classical method of multiplication gives

$$M(n) = O(n^2)$$

By splitting the factors in two $n/2$-digit numbers and using

$$a = a_1 B^{n/2} + a_2, \qquad b = b_1 B^{n/2} + b_2$$

$$p_1 = a_1 b_1, \qquad p_2 = a_2 b_2, \qquad p_3 = (a_1 + a_2)(b_1 + b_2)$$

$$ab = p_1 B^n + (p_3 - p_2 - p_1) B^{n/2} + p_2$$

we obtain

$$M(n) = 3M(n/2) + O(n) = O(n^{1.58496\ldots})$$

Similarly, by splitting the numbers in k (n/k)-digit components,

$$M(kn) = (2k-1)M(n) + O(n) = O(n^{\log_k (2k-1)})$$

By the application of a technique resembling the Fast Fourier Transform and modular arithmetic,

$$M(n) = O(n \log(n) \log(\log n))$$

Note that the complexity of multiplication is bounded above by the complexity of squaring and by the complexity of computing inverses. I.e.

$$Q_{\times}(n) \le 2Q_{x^2}(n) + O(n)$$

since

$$ab = \frac{(a+b)^2 - (a-b)^2}{4}$$

and

$$Q_{x^2}(n) \le 3Q_{1/x}(n) + O(n)$$

since

$$x^2 = \frac{1}{\frac{1}{x-1} - \frac{1}{x}} + x$$

For the next complexity results we will assume that we use an asymptotically fast multiplication algorithm, i.e. one for which

$$M(n) = O(n(\log n)^k)$$

In such circumstances,

$$\sum_{k \ge 0} M(n\alpha^k) = \frac{M(n)}{1-\alpha}\left(1 + O(1/(\log n))\right)$$

Inverses ($x = 1/a$) can be computed using variable precision steps with the second order iterative formula:

$$x_{i+1} = x_i(2 - ax_i)$$

Each step requires two multiplications and one addition. Since this Newton-type iteration converges quadratically, the last iteration is done with n digits, the previous to the last with $\lceil n/2 \rceil$, the previous with $\lceil n/4 \rceil$, etc.

$$\begin{aligned} Q_{1/x}(n) &= \sum_{i \ge 0} 2M(\lceil n/2^i \rceil) + O(n/2^i) \\ &\approx 3M(n) \end{aligned}$$

If we use a third order method:

$$x_{i+1} = x_i(\varepsilon_i^2 - \varepsilon_i + 1) \qquad \varepsilon_i = ax_i - 1$$

then $Q_{1/x}(n) \approx 3M(n)$ also. Consequently divisions can be computed in

$$Q_{/}(n) \approx 4M(n)$$

To evaluate $x = a^{-1/2}$ we can use the third-order iteration:

$$\varepsilon_i = a x_i^2 - 1$$

$$x_{i+1} = x_i - x_i \varepsilon_i \frac{4 - 3\varepsilon_i}{8}$$

for which

$$Q_{1/\sqrt{x}}(n) \approx \frac{9M(n)}{2}$$

Consequently

$$Q_{\sqrt{x}}(n) \approx \frac{11M(n)}{2}$$

Derivatives can be computed from the formula

$$f'(x) = \frac{f(x+h) - f(x-h)}{2h} + O(h^3)$$

by making $h = O(f'(x)B^{-n/3})$. For this method

$$Q_{f'(x)}(n) = 2Q_{f(x)}(3n/2) + O(n)$$

The inverse of a function can be computed by using any iterative zero-finder with variable precision. By using the secant method:

$$x_{i+1} = x_i - f(x_i)\frac{x_i - x_{i-1}}{f(x_i) - f(x_{i-1})}$$

then

$$Q_{f^{-1}(x)}(n) \approx 15M(n) + Q_{f(x)}(n) + \sum_{i \geq 2} Q_{f(x)}(2n\rho^{-i})$$

where $\rho = (1+\sqrt{5})/2$ is the golden ratio.

For the purpose of describing the algorithms we will use a common representation, based on arrays of digits. The digits may take values from 0 to $BASE - 1$ in their normalized form, although a digit may hold a maximum value $MAXD$. For example, for eight-bits characters on which we want to represent decimal numbers, $BASE = 10$ and $MAXD = 255$. The bound $MAXD$ may be any value including $BASE - 1$. For our algorithms we will assume that $MAXD \geq 2BASE^2$. With this assumption we do not have to use temporary variables for the handling of digits.

The data definition for our C algorithms is

```
typedef digit mp[];
```

---

mp[0] will be called the *header* and will be used to store control information about the number. Typical control information is sign, length, exponent (for floating point implementations), etc. We are not concerned about the organization of bits in the header, as long as we can store and retrieve its values. The lowest order digit is stored in mp[1]; the highest order digit is stored in mp[length(mp)-1]. This organization, although not very common, is quite convenient.

The following procedure normalizes a multiple precision number, adjusts its length, propagates carries and adjusts sign if needed.

Normalization of a multiple-precision number

```
normalize( a )
mp a;

{int cy, i, la;
la = length(a);
start:
cy = 0;
for ( i=1; i<la; i++ )     {
        cy = (a[i] += cy) / BASE;
        a[i] -= cy*BASE;
        if (a[i]<0)  {a[i] += BASE;    cy--;}
        }
while (cy>0) {
        a[i++] = cy%BASE;
        cy /= BASE;}
if (cy<0)     {
        a[la-1] += cy*BASE;
        for (i=1; i<la; i++)      a[i] = -a[i];
        storesign( a, sign(a)==POS ? NEG : POS );
        goto start;
        }
while (a[i-1]==0 && i>2) i--;
storelength( a, i );
if ( i==2 && a[1]==0 )  storesign( a, POS );
};
```

The following procedure computes a linear combination of two multiple-precision numbers. The integer coefficients should be in the range $-BASE...BASE$. The result is computed, destructively, on the first argument.

Linear combination of two numbers

```
linear( a, ka, b, kb )
mp a, b;
int ka, kb;

/*** compute a*ka + b*kb --> a ***/
{int i, la, lb;
la = length( a );    lb = length( b );
for (i=1; i<la; i++) a[i] *= ka;
if ( sign(a)≠sign(b) )       kb = -kb;
if ( lb>la ) {
        storelength( a, lb );
        for (i=la; i<lb; i++) a[i] = 0;
        }
for (i=1; i<lb; i++) a[i] += kb*b[i];
normalize( a );
};
```

Multiple-precision multiplication

```
mulint( a, b, c )
mp a, b, c;
/* Multiply two integers.  a*b-->c */

{int i, j, la, lb;
/* b and c may coincide */
la = length(a); lb = length(b);
for (i=0; i<la-2; i++)   c[lb+i] = 0;
for (i=lb-1; i>0; i--)   {
        for (j=2; j<la; j++)
                if ( (c[i+j-1] += b[i]*a[j]) > MAXD-BASE*BASE ){
                        c[i+j-1] -= (MAXD/BASE)*BASE;
                        c[i+j] += MAXD/BASE;
                        }
        c[i] = b[i]*a[1];
        }
storelength( c, la+lb-2 );
storesign( c, sign(a)==sign(b) ? POS : NEG );
normalize( c );
};
```

References:
[Aho *et al.*, 74], [Alt, 80], [Alt, 83], [Borodin *et al.*, 75], [Brent, 76], [Collins *et al.*, 77], [Dhawan *et al.*, 77], [Flajolet *et al.*, 83], [Floyd, 75], [Head, 80], [Ja'Ja', 79], [Knuth, 69], [Knuth, 78], [Morris, 78], [Stockmeyer, 83].

6.2 Other arithmetic functions

6.2.1 Binary powering

Binary powering is a tail recursion technique for powering a number. To compute a given power, we first compute the power to half the exponent and then square the result. If the exponent is odd, we additionally multiply the result by the base. Let $Q_{bp}(n)$ denote the number of multiplications required to compute the nth power of a number using binary powering. Then

$$Q_{bp}(n) \; = \; \lfloor \log_2 n \rfloor + \nu(n) - 1$$

where $\nu(n)$ is the number of "one" digits in the binary representation of n.

Let $Q_{opt}(n)$ be the number of multiplications required by the optimal method of powering, i.e. the method which minimizes the number of multiplications. Then

$$1 \; \leq \; \frac{Q_{opt}(n)}{\lfloor \log_2 n \rfloor} \; \leq \; 1 + \frac{1}{\log_2 \log_2 n} + O\left(\frac{\log_2 \log_2 \log_2 n}{(\log_2 \log_2 n)^2} \right)$$

and

$$Q_{opt}(n) \; \leq \; Q_{bp}(n)$$

The first inequality is tight, but the latter is not. $n=15$ is the smallest example for which they differ: we can compute x^{15} by computing x^2, x^3, x^6, x^{12} and x^{15} giving $Q_{opt}(15)=5$ while $Q_{bp}(15)=6$. Similarly, the smallest exponent for which the difference is 2 is 63, $Q_{opt}(63)=8$ while $Q_{bp}(63)=10$. (One of the optimal sequences of powers is 2,4,5,9,18,27,45,63.)

The problem of computing the optimal strategy for powering is related to the *addition chain* problem, i.e. how to construct an increasing sequence $a_1, a_2, \ldots, a_k$ for which every element is the sum of two previous elements and $a_1=1$ and $a_k=n$ for a minimal k.

Using the fact that $(a^x)^y = a^{xy}$, if the power is a composite number, then

$$Q_{opt}(pq) \; \leq \; Q_{opt}(p) + Q_{opt}(q)$$

This inequality is not tight. For example, $Q_{opt}(33)=6$ but $Q_{opt}(3)=2$ and $Q_{opt}(11)=5$.

It is always possible to do a squaring as the last step, which gives

$$Q_{opt}(2n) \; \leq \; Q_{opt}(n) + 1$$

but this bound is not tight either since $Q_{opt}(191)=11$ and $Q_{opt}(382)=11$.

For binary powering we can define an average value of the complexity, as if the bits of the power were randomly selected. For this definition

$$\overline{Q_{bp}(n)} = \sum_{2^k}^{2^{k+1}-1} Q_{bp}(k) = \frac{3k}{2}$$

where $k = \lfloor \log_2 n \rfloor$.

When powering integers, as the powers grow in size, it is important to know the complexity of the multiplication method used. Let n denote the exponent and N the number of digits in the base number to be powered. If we use the classical algorithm, $M(N)=O(N^2)$ then

$$Q_{bp}(n) = \left(\frac{n^2}{3} + O(n)\right) M(N)$$

The iterative version of the powering algorithm runs in the order

$$Q_{iter}(n) = \frac{n(n-1)}{2} M(N)$$

If we use an asymptotically fast multiplication algorithm, $(M(N)=O(N(\log N)^k))$, then binary powering is definitely better than iterative powering:

$$Q_{bp}(n) \approx 2M(Nn)$$

as opposed to

$$Q_{iter}(n) \approx \frac{n^2}{2} M(N)$$

In the above cases it is assumed that the size of the result of powering an N-digit number to the nth power is an Nn-digit number. This may be too pessimistic sometimes.

Binary powering

```
function power( b : number; e : integer ) : number;
begin
if e<0 then  power := 1/power(b,-e)
else if e=0 then      power := 1
else if e=1 then      power := b
else if (e mod 2) = 0 then
                      power := sqr( power( b, e div 2 ) )
else                  power := sqr( power( b, e div 2 ) ) * b
end;
```

6.2.2 Arithmetic-geometric mean

The Arithmetic-geometric mean (AG mean) constructs two sequences of numbers a_i and b_i from starting values a_0 and b_0 and the iteration formulas:

$$a_{i+1} = \frac{a_i + b_i}{2}$$

$$b_{i+1} = \sqrt{a_i b_i}$$

For $0 < a_0 \leq 1$ and $0 < b_0 \leq 1$ the sequences converge quadratically to their common limit denoted by $AG(a_0, b_0)$. Computing one step of the iteration requires one multiplication and one square root plus other $O(n)$ operations. Consequently the complexity of the AG computation is

$$Q_{AG}(n) \approx \frac{13}{2} M(n) \log_2 n$$

where n is the number of digits in the answer. The AG mean is related to the complete elliptic integrals as

$$\frac{\pi}{2AG(1, \cos\phi)} = \int_0^{\pi/2} \frac{d\theta}{\sqrt{1 - \sin^2\phi \sin^2\theta}}$$

The *Brent-Salamin* method for computing π which uses the AG mean and a Legendre's identity requires

$$Q_\pi(n) \approx \frac{15}{2} M(n) \log_2 n$$

Fast computation of π

```
function pi : number;
var a, b, t, x, tempa : number;

begin
a := 1;
b := sqrt( 0.5 );
t := 0.25;
x := 1;
while a-b>epsilon do begin
        tempa := a;
        a := (a+b) / 2;
        b := sqrt( tempa*b );
        t := t - x*sqr(a-tempa);
        x := 2*x
        end;
pi := sqr(a+b) / (4*t)
end;
```

Other classical methods for evaluating π are based on identities of the type

$$\pi = 16\tan^{-1}(1/5) - 4\tan^{-1}(1/239)$$

The function $\tan^{-1}(1/i)$ for integer i can be evaluated in time proportional to $O(n^2/\log i)$ using the Maclaurin expansion of $\tan^{-1}(x) = x - x^3/3 + x^5/5 \ldots$

6.2.3 Transcendental functions

Assuming that π and $\ln(BASE)$ are precomputed to the desired accuracy, we can compute $\ln(x)$ using the formula

$$\ln(x) = \frac{\pi}{2AG(1,4/x)}(1+O(x^{-2}))$$

If x is not large enough, we can simply scale it by multiplying by a suitable power of the $BASE$ (just a shift). For this method

$$Q_{\ln(x)}(n) \approx 13M(n)\log_2 n$$

Computation of natural logarithms by AG means

```
function ln ( x : number ) : number;
var     a, b, temp : number;
        shift, logbase : integer;

begin
logbase := crude-estimate-of-ln(x)/ln(BASE);
if 2*logbase<Digits then begin
        shift := Digits div 2 - logbase + 1;
        ln := ln( x * BASE**shift ) - shift*LNBASE
        end
else    begin
        a := 1.0;     b := 4/x;
        while a-b>sqrteps do begin
                temp := a;
                a := (a+b) / 2;
                b := sqrt( temp*b )
                end;
        ln := Pi / (a+b)
        end
end;
```

The above algorithm uses two pre-computed constants: *Pi* and *LNBASE*, with their obvious meanings. *LNBASE* can be computed with the above function by computing the logarithm of $BASE^{Digits}$ and then dividing the result by *Digits*. The global variable *Digits* indicates the precision of the computation, or how many significant digits in base *BASE* are kept. *epsilon* is a bound on the desired error and *sqrteps* is the square root of epsilon.

By computing inverses with a method of high order of convergence (in this case all the derivatives are easy to compute) we obtain

$$Q_{e^x}(n) \approx 13M(n)\log_2 n$$

By doing all the arithmetic operations with complex numbers, or by computing the $\tan^{-1}(x)$ as below, we can compute all the trigonometric functions and their inverses. E.g.

$$\sin x = \frac{e^{ix} - e^{-ix}}{2i}$$

Then

$$Q_{\sin x}(n) \approx Q_{\cos x}(n) \approx \cdots \approx Q_{\sin^{-1} x}(n) \approx \cdots \approx 34M(n)\log_2 n$$

Computation of arctan(x) by AG means

```
function arctan ( x : number ) : number;
var    q, s, v, w : number;

begin
s := sqrteps;
v := x / (1 + sqrt(1+x*x));
q := 1;
while 1-s > epsilon do begin
        q := 2*q / (1+s);
        w := 2*s*v / (1+v*v);
        w := w / (1 + sqrt(1-w*w));
        w := (v+w) / (1-v*w);
        v := w / (1 + sqrt(1+w*w));
        s := 2*sqrt(s) / (1+s)
        end;
arctan := q * ln( (1+v)/(1-v) )
end;
```

References:
[Borodin *et al.*, 75], [Brent, 76], [Downey *et al.*, 81], [Horowitz, 73], [Kedem, 74], [Pippenger, 76], [Pippenger, 79], [Pippenger, 80], [Winograd, 75], [Yao, 76].

6.3 Matrix multiplication

For any matrices

```
a : array [1..m, 1..p ] of basetype;
b : array [1..p, 1..n ] of basetype;
c : array [1..m, 1..n ] of basetype;
```

---

we define the matrix product $c = a \times b$ by

$$c_{ij} = \sum_{k=1}^{p} a_{ik} b_{kj}$$

The classical algorithm for matrix multiplication requires mnp multiplications and $mn(p-1)$ additions. Let $M_{\times}(n)$ be the number of multiplications used to multiply two $n \times n$ matrices. Then $M_{\times}(n) = n^3$ for the classical algorithm.

Classical

```
for i:=1 to m do
        for j:=1 to n do begin
                c[i,j] := 0;
                for k:=1 to p do
                        c[i,j] := c[i,j] + a[i,k]*b[k,j]
                end;
```

Winograd's method of general matrix multiplication reduces the number of multiplication to about half with the formula:

$$c_{ij} = \sum_{k=1}^{p/2} (a_{i,2k} + b_{2k-1,j})(a_{i,2k-1} + b_{2k,j}) - d_i - e_j + a_{ip} b_{pn} \dagger$$

where

$$d_i = \sum_{k=1}^{p/2} a_{i,2k} a_{i,2k-1}$$

$$e_j = \sum_{k=1}^{p/2} b_{2k-1,j} b_{2k,j}$$

and the last term (†) is present only if p is odd. Winograd's matrix multiplication uses

$$M_{\times}(m,p,n) = mn\lceil p/2\rceil + (m+n)\lfloor p/2\rfloor$$

multiplications and

$$A_{\times}(m,p,n) = mn(p+2) + (mn+m+n)(\lfloor p/2\rfloor - 1)$$

additions/subtractions.

6.3.1 Strassen's matrix multiplication

When $m = n = p = 2$, the product can be computed using 7 multiplications instead of 8 but using 15 additions instead of 4.

$$\begin{array}{rclrclrcl}
s_1 &=& a_{21}+a_{22} & p_1 &=& s_2 s_6 & s_8 &=& s_6-b_{21} \\
s_2 &=& s_1-a_{11} & p_2 &=& a_{11}b_{11} & s_9 &=& p_1+p_2 \\
s_3 &=& a_{11}-a_{21} & p_3 &=& a_{12}b_{21} & s_{10} &=& s_9+p_4 \\
s_4 &=& a_{12}-s_2 & p_4 &=& s_3 s_7 & c_{11} &=& p_2+p_3 \\
s_5 &=& b_{12}-b_{11} & p_5 &=& s_1 s_5 & c_{12} &=& s_9+p_5+p_6 \\
s_6 &=& b_{22}-s_5 & p_6 &=& s_4 b_{22} & c_{21} &=& s_{10}-p_7 \\
s_7 &=& b_{22}-b_{12} & p_7 &=& a_{22}s_8 & c_{22} &=& s_{10}+p_5
\end{array}$$

This can be applied not only to 2×2 matrices, but to any matrix $n\times n$, partitioned into 4 $(n/2)\times(n/2)$ matrices (with proper 0 padding if necessary). The number of multiplications required by a recursive application of Strassen's algorithm to multiply two $2^k\times 2^k$ matrices is

$$M_{\times}(2^k) = 7^k$$

and in general

$$M_{\times}(n) = 7M_{\times}(\lceil n/2\rceil) = O(n^{2.80735...})$$

Let $A_{\times}(n)$ be the number of additions used to multiply two $n\times n$ matrices, then

$$A_{\times}(2^k) = 5(7^k - 4^k)$$

$$A_{\times}(n) = 7A_{\times}(\lceil n/2\rceil) + 15\lceil n/2\rceil^2 = O(n^{2.80735...})$$

For the implementation of this algorithm we are interested in the total number of additions/ multiplications. Noticing that when n is odd, one of the recursive matrix multiplications can be done on matrices $\lfloor n/2\rfloor\times\lfloor n/2\rfloor$ and by shifting to the classical algorithm whenever it is more efficient we obtain that the total number of operations is

$$\begin{aligned} M_{\times}(n) &= \min(M_{\times}(\lfloor n/2\rfloor) + 6M_{\times}(\lceil n/2\rceil) + 15\lceil n/2\rceil^2,\ n^2(2n-1)) \\ &\approx 3.73177...n^{2.80735...} \end{aligned}$$

Even for this optimized implementation, n has to be larger than 1580 to save 50% of the operations with respect to the classical method.

6.3.2 Further asymptotic improvements

The following methods present asymptotic improvements to the number of operations necessary to do matrix multiplication. These improvements are only of theoretical interest, as their complexity for normal size problems is much worse than the classical algorithm. Furthermore, their numerical properties are unknown.

Pan devised a general multiplication scheme using trilinear forms which requires $n^3/3+6n^2-4n/3$ multiplications to multiply to $n \times n$ matrices. His method does not rely on product commutativity and can be composed as we did with Strassen's. By selecting as a base matrices with 70×70 we obtain

$$M_{\times}(n) \;=\; O(n^{\omega})$$

where $\omega = \dfrac{\ln 143640}{\ln 70} = 2.79512...$

Bini *et al.* use an approximate (arbitrary precision approximating) method to multiply 12×12 matrices with 1000 multiplications and hence, extending their method gives

$$\omega \;=\; \frac{\ln 1000}{\ln 12} \;=\; 2.77988...$$

Schonhage generalized the above method to obtain

$$\omega \;=\; 2.54799...$$

where $16^{\omega/3} + 9^{\omega/3} = 17$. Pan further improved this with a construction that achieves

$$\omega \;=\; \frac{3\ln 52}{\ln 110} \;=\; 2.52181...$$

Coppersmith *et al.* describe the construction of a faster matrix multiplication algorithm based on any given algorithm using bilinear forms. Using the above they construct a method for which $\omega < 2.49554....$

Matrix inversion, computation of determinants, solution of simultaneous equations and lower-upper triangular factoring can be done in terms of matrix multiplications and hence can be done in time proportional to matrix multiplication.

References:
[Aho *et al.*, 74], [Bini *et al.*, 79], [Borodin *et al.*, 75], [Brockett *et al.*, 76], [Cohen *et al.*, 76], [Cohen, 83], [Coppersmith *et al.*, 82], [Coppersmith, 82], [Dobkin *et al.*, 76], [Dobkin, 73], [Feig, 81], [Feig, 83], [Hu *et al.*, 82], [Ja'Ja', 79], [Ja'Ja', 80], [Ja'Ja', 83], [Knuth, 69], [Lotti *et al.*, 80], [Makarov, 81], [Pan, 78], [Pan, 79], [Pan, 80], [Pan, 81], [Pan, 81], [Pan, 83], [Probert, 76], [Probert, 76], [Probert, 78], [Romani, 82], [Santoro, 80], [Savage, 74], [Schachtel, 78], [Schonhage, 81], [Schoor, 82], [Strassen, 69], [Winograd, 68], [Yuval, 78].

6.4 Polynomial evaluation

The simplest method of evaluating a polynomial is with Horner's rule. Let

$$P(x) \;=\; a_0 + a_1 x + a_2 x^2 + \cdots + a_n x^n$$

then, $P(x)$ can be expressed as

$$P(x) \;=\; a_0 + x(a_1 + x(a_2 + x(\cdots x a_n) \cdots)))$$

Let

---

a : **array** [0..*n*] **of** *basetype;*

---

be the array containing the coefficients of an nth degree polynomial. Then

Horner's polynomial evaluation

```
result := a[n];
for i:=n-1 downto 0 do
        result := result * x + a[i];
```

For evaluating a polynomial at a single point, Horner's rule is optimal with respect to the number of additions and multiplications. In general this method has very good numerical properties too.

If the same polynomial is evaluated at several points, then it is possible to do some set up work in order to save time during the evaluation. This initial task is called *preconditioning*. It is normally assumed that the cost of preconditioning is not significant, i.e. the polynomial will be evaluated so many times that the fixed initial cost can be discarded.

Table 6.1 shows upper bounds for evaluating polynomials over the real and over the complex numbers. The bounds are shown as pairs (m,a), where m indicates the number of multiplications and a indicates the number of additions.

† indicates an additional shifting operation. Shifting the radix point can be considered as being of the same complexity as addition.

Table 6.2 shows lower bounds on the number of additions and multiplications for polynomial evaluation with preconditioning. Unless one of the values is missing, these bounds are simultaneous lower bounds; that is no algorithm can perform polynomial evaluation with less than m multiplications and less than a additions.

degree	Reals	Complex	References
4,6	$(\frac{n+2}{2}, n+1)$	same	[Motzkin,55], [Knuth,62]
any	$(\lceil\frac{n+3}{2}\rceil, n)$	same	[Eve,64], [Knuth,62]
even, $n \geq 8$	$(\frac{n+2}{2}, n+3$ *or* $n+1\dagger)$	$(\frac{n+2}{2}, n)$	[Pan,79], [Eve,64]
odd, $n \geq 11$	$(\frac{n+1}{2}, n+4\dagger)$	$(\frac{n+1}{2}, n+2)$	[Pan,79]
9		$(\frac{n+1}{2}, n+3)$	[Revah,75]

Table 6.1: Upper bounds for polynomial evaluation with preconditioning

degree	$(\times, +)$	References
$n \geq 2$	$(\lceil\frac{n+1}{2}\rceil, ...)$	[Motzkin,55]
odd, $n \geq 7$	$(\frac{n+3}{2}, ...)$	[Motzkin,55], [Knuth,81], [Revah,75]
any	$(..., n)$	[Belaga,58]
4,6,8	$(\frac{n+2}{2}, n+1)$	[Knuth,81], [Pan,79]
odd, $n \geq 11$	$(\frac{n+1}{2}, n+2)$	[Knuth,62], [Revah,75]
odd, $n \geq 3$	$(\frac{n+3}{3}, n)$	[Belaga,58], [Revah,75]

Table 6.2: Lower bounds for polynomial evaluation with preconditioning

References:

[Aho *et al.*, 74], [Aho *et al.*, 75], [Belaga, 58], [Borodin *et al.*, 75], [Borodin *et al.*, 76], [Brown *et al.*, 80], [Chin, 76], [Dobkin *et al.*, 80], [Eve, 74], [Heintz *et al.*, 80], [Heintz *et al.*, 80], [Horowitz, 73], [Horowitz, 74], [Hyafil *et al.*, 75], [Knuth, 62], [Knuth, 69], [Kung, 73], [Lipton *et al.*, 75], [Lipton *et al.*, 76], [Lipton, 78], [Moenk *et al.*, 72], [Motzkin, 65], [Nozaki, 79], [Pan, 78], [Pan, 79], [Pan, 81], [Revah, 75], [Rivest *et al.*, 79], [Savage, 74], [Schnorr *et al.*, 80], [Schnorr, 81], [Shaw *et al.*, 74], [Shaw *et al.*, 77], [Strassen, 74], [Wiele, 78].

Appendix I: DISTRIBUTIONS DERIVED FROM EMPIRICAL OBSERVATION

In this appendix we will describe some probability distributions arising from empirical situations. The distributions described here may be used with other well known distributions to test algorithms under various conditions. Some of these distributions are related directly to data processing.

I.1 Zipf's law

Zipf observed that the frequency of word usage (in written English) follows a simple pattern. When word frequencies are listed in decreasing order, we have the relation

$$f_1 \approx i f_i$$

where f_i denotes the frequency of the ith most frequent word. Zipf observed that the population of cities in the U.S. also follows this relation closely. From this observation we can easily define a Zipfian probability distribution as

$$p_i = \frac{1}{iH_n} \qquad 1 \le i \le n$$

The first moments and variance of this distribution are

$$\mu_1' = \frac{n}{H_n}$$

$$\mu_2' = \frac{n(n+1)}{2H_n}$$

$$\sigma^2 = \frac{n}{H_n}\left(\frac{n+1}{2} - \frac{n}{H_n}\right)$$

This distribution can be generalized in the two following ways.

I.1.1 First generalization of a Zipfian distribution

In this case the probabilities are defined by

$$p_i = \frac{1}{a(i+b)} \qquad (1 \le i \le n,\ b > -1)$$

where $a = \psi(n+b+1) - \psi(b+1)$. The first moments and variance are

$$\mu_1' = \frac{n}{a} - b$$

$$\mu_2' = \frac{n(n+1) - 2nb + 2ab^2}{2a}$$

$$\sigma^2 = \frac{n}{2a}(n+1+2b-2n/a)$$

Choosing b to be an integer allows us to represent truncated Zipfian distributions. Giving b a small non-integer value may provide a better fit for the first few frequencies.

I.1.2 Second generalization of a Zipfian distribution

This generalization introduces a parameter θ so that we may define

$$p_i = \frac{1}{i^\theta H_n^{(\theta)}} \qquad 0 \le \theta < 1$$

Zipf found that some word frequencies matched this distribution closely for values of θ other than 1. In this case the first moments and variance are

$$\mu_1' = \frac{H_n^{(\theta-1)}}{H_n^{(\theta)}} = \frac{n(1-\theta)}{2-\theta} + O(n^\theta)$$

$$\mu_2' = \frac{H_n^{(\theta-2)}}{H_n^{(\theta)}} = \frac{n^2(1-\theta)}{3-\theta} + O(n^{1+\theta})$$

$$\sigma^2 = \frac{H_n^{(\theta-2)} H_n^{(\theta)} - (H_n^{(\theta-1)})^2}{(H_n^{(\theta)})^2} = \frac{n^2(1-\theta)}{(3-\theta)(2-\theta)^2} + O(n^{1+\theta})$$

References:
[Johnson *et al.*, 69], [Knuth, 73], [Zipf, 49].

I.2 Bradford's law

Bradford's law was first observed in experiments dealing with the number of references made to a selection of books in search of information. This principle can be described in the following way. Assume that we have a collection of n books which treat a given topic, and that these books are placed on a shelf in decreasing order according to the number of times each book is referenced. Thus the most referenced book is first and so on. We then divide these books into k contiguous groups such that each group receives the same number of references. Bradford's law now states that the number of books in each successive division follows the ratio $1{:}m{:}m^2{:} \cdots {:}m^{k-1}$ for some constant m.

To translate this description into mathematical terms, we let r_i be the expected value of the number of references to the ith most referenced book on our shelf. Thus we have $r_1 \ge r_2 \ge \cdots \ge r_n$. Let $R(j)$ be the partial sum of the expected values of these references:

$$R(j) = \sum_{i=1}^{j} r_i$$

and so

$$R(n) = T$$

where T is the total expected number of references. To divide the n books into k divisions satisfying the given ratio, the number of books in each division must be $\frac{n(m-1)}{m^k-1}, \frac{nm(m-1)}{m^k-1}, \ldots, \frac{nm^{k-1}(m-1)}{m^k-1}$. Since each division receives the same number of references, this number must be T/k. Consequently the total expected number of references to the first division will be

$$\sum_{i=1}^{\frac{n(m-1)}{m^k-1}} r_i = R\left(\frac{n(m-1)}{m^k-1}\right) = \frac{T}{k}$$

In general, for the first j divisions we have the equation

$$R\left(\frac{(m^j-1)n}{m^k-1}\right) = \frac{jT}{k} \qquad (1)$$

Now the quantities k and m are related to one another, since for any valid k, Bradford's law predicts the existence of a unique m. Examination of $R(x)$ for different values of k and m shows that in order for the law to be consistent, the quantity $m^k-1 = b$ must be constant. This constant b defines the shape of the distribution. From equation (1) we can solve for $R(x)$ and obtain

$$R(x) = \frac{T}{k}\log_m\left(\frac{bx}{n}+1\right)$$

Let p_i be the probability that a random reference refers to the ith book. From the above discussion we have

$$p_i = \frac{R(i)-R(i-1)}{T} = \frac{1}{k}\log_m\left(\frac{bi+n}{b(i-1)+n}\right)$$

Since $m^k-1 = b$, we have $k\ln m = \ln(b+1)$; this allows us to simplify the given probability to

$$p_i = \log_{b+1}\left(\frac{bi+n}{b(i-1)+n}\right)$$

The first moment of the above distribution is

$$\mu_1' = \sum_{i=1}^{n} ip_i = n\log_{b+1}\left(\frac{n(b+1)}{b}\right) - \log_{b+1}\left(\frac{\Gamma(n(b+1)/b)}{\Gamma(n/b)}\right) \qquad (2)$$

$$= n\left(\frac{1}{\ln(b+1)} - \frac{1}{b}\right) + \frac{1}{2} + \frac{b^2}{12n(b+1)\ln(b+1)} + O(n^{-3})$$

The second moment is given by

$$\mu_2' = \frac{n^2}{b^2} - \frac{n}{b} + \frac{1}{3} + \frac{1}{\ln(b+1)}\left(\frac{n^2(b-2)}{2b} + n - \frac{b}{6(b+1)} + \frac{b^2}{12(b+1)n}\right) + O(n^{-2}) \quad (3)$$

The variance is

$$\sigma^2 = \frac{n^2}{\ln(b+1)}\left(\frac{b+2}{2b} - \frac{1}{\ln(b+1)}\right) + O(1) \quad (4)$$

This distribution behaves very much like the generalized harmonic (or the first generalization of Zipf's distribution). When the parameter $b \rightarrow 0$ Bradford's distribution coincides with the discrete rectangular distribution.

Although the process of accessing information from books is rarely automated, there is a significant number of automatic processes in which the accessing of information is similar to the situation of referencing books. In these cases Bradford's law may provide a good model of the access probabilities.

References:

[Pope, 75].

I.3 Lotka's law

Lotka observed that the number of papers in a given journal written by the same author closely followed an inverse square distribution. In other words, if we were to choose an author at random from the list of contributors to the journal, the probability that he or she had contributed exactly i papers would be proportional to i^{-2}. Later it was observed that for some journals an inverse cube law fit the data more precisely. We will generalize these two laws in the following way. Let n be the total number of authors who published at least one paper in a given journal. The probability that a randomly chosen author contributed exactly i papers will be given by

$$p_i = \frac{1}{\zeta(\theta) i^\theta}$$

The first moment of this distribution corresponds to the average number of papers published by each author; it is given by

$$\mu_1' = \sum_{i=1}^{\infty} i p_i = \frac{\zeta(\theta-1)}{\zeta(\theta)}$$

We immediately conclude that this law will be consistent only for $\theta > 2$, as has been noted by several other authors; otherwise this first moment will be unbounded, a situation which does not correspond with reality. Note that $n\mu'_1$ denotes the expected number of papers published in a journal which has

n contributors.

For $\theta \leq 3$, the variance of the distribution under discussion diverges. For $\theta > 3$, the variance is given by

$$\sigma^2 = \frac{\zeta(\theta-2)}{\zeta(\theta)} - \left(\frac{\zeta(\theta-1)}{\zeta(\theta)}\right)^2$$

The median number of papers by the most prolix author can be approximated by

$$median \approx \left(\frac{n}{\ln(2)\zeta(\theta)(\theta-1)}\right)^{\frac{1}{\theta-1}}$$

References:
[Lotka, 26], [Murphy, 73], [Radhakrishnan *et al.*, 79].

I.4 80%-20% rule

The 80%-20% rule was proposed as a probabilistic model to explain certain data processing phenomena. In computing folklore it is usually given as: "80% of the transactions are on the most active 20% of the records, and so on recursively." Mathematically, let $p_1 \geq p_2 \geq p_3 \geq \cdots \geq p_n$ be the independent probabilities of performing a transaction on each of the n records. Let $R(j)$ be the cumulative distribution of the p_is, i.e.

$$\sum_{i=1}^{j} p_i = R(j); \qquad R(n) = 1$$

The 80%-20% rule is expressed in terms of the function $R(j)$ by

$$R(n \times 20\%) = 80\%$$

This rule may be applied recursively by requiring that the relation hold for any contiguous subset of p_is that includes p_1. This requirement yields the necessary condition:

$$R(0.2j) = 0.8R(j)$$

More generally we may consider an $\alpha\% - (1-\alpha)\%$ rule given by

$$R((1-\alpha)j) = \alpha R(j), \qquad \frac{1}{2} \leq \alpha \leq 1 \qquad (1)$$

The above functional equation defines infinitely many probability distributions for each choice of α. One simple solution that is valid for all real j is

$$R(i) = \frac{i^\theta}{n^\theta}$$

where $\theta = \dfrac{\ln(\alpha)}{\ln(1-\alpha)}$. Thus $0 \leq \theta \leq 1$. This formula for $R(i)$ implies

$$p_i = \frac{i^\theta - (i-1)^\theta}{n^\theta} \qquad (2)$$

Note that this probability distribution also possesses the required monotone behaviour, i.e. $p_i \geq p_{i+1}$.

The parameter θ gives shape to the distribution. When $\theta = 1$ $(\alpha = \frac{1}{2})$ the distribution coincides with the discrete rectangular distribution. The moments and variance of the distribution described by equation (2) are

$$\mu_1' = \sum_{i=1}^{n} i p_i = \frac{\theta n}{\theta+1} + \frac{1}{2} - \frac{\zeta(-\theta)}{n^\theta} - \frac{\theta}{12n} + O(n^{-3})$$

$$\mu_2' = \sum_{i=1}^{n} i^2 p_i = \frac{\theta n^2}{\theta+2} + \frac{\theta n}{\theta+1} + \frac{2-\theta}{6} - \frac{2\zeta(-\theta-1)+\zeta(-\theta)}{n^\theta} + O(n^{-1})$$

$$\mu_3' = \sum_{i=1}^{n} i^3 p_i = \frac{\theta n^3}{\theta+3} + \frac{3\theta n^2}{2(\theta+2)} + \frac{\theta(3-\theta)n}{4(\theta+1)} + O(1)$$

$$\mu_k' = \sum_{i=1}^{n} i^k p_i = \frac{\theta n^k}{\theta+k} + \frac{\theta k n^{k-1}}{2(\theta+k-1)} + \frac{\theta(k-\theta)kn^{k-2}}{12(\theta+k-2)} + O(n^{k-3}) + O(n^{-\theta})$$

$$\sigma^2 = \frac{\theta n^2}{(\theta+1)^2(\theta+2)} + O(n^{1-\theta})$$

For large n, the tail of the distribution coincides asymptotically with $p_i \approx i^{\theta-1}$. For the 80%-20% rule, $\theta = 0.138646...$; consequently the distribution which arises from this rule behaves very similarly to the second generalization of Zipf's distribution.

References:

[Heising, 63], [Knuth, 73].

Appendix II: ASYMPTOTIC EXPANSIONS

This appendix contains a collection of asymptotic expansions of functions or expressions commonly used in the analysis of algorithms. The criteria used for the length of the expansion i.e. order, is rather artificial and depends upon computability and number of terms in the numerator, and is at most 7.

It is assumed that the expansions are for $n \to \infty$ unless otherwise specified. It is also assumed that a, b, c and z are all $O(1)$ when $n \to \infty$.

In the following, $\zeta(z)$ is the classical Riemann zeta function, defined by

$$\zeta(z) = \sum_{n=1}^{\infty} n^{-z}$$

$\Gamma(z)$ denotes the Gamma function, defined by

$$\Gamma(z+1) = \int_0^{\infty} t^z e^{-t} dt$$

$$= z\Gamma(z)$$

$\psi(z)$ denotes the Psi function, defined by

$$\psi(z+1) = \frac{\Gamma'(z+1)}{\Gamma(z+1)} = \psi(z) + \frac{1}{z}$$

and γ will denote Euler's constant,

$$\gamma = \lim_{n \to \infty} H_n - \ln(n) = 0.5772156649...$$

II.1 Asymptotic expansions of sums

$$\sum_{k=1}^{n-1} [k(n-k)]^{-1/2} = \pi + n^{-1/2}\left(2\zeta(1/2) + \frac{\zeta(-1/2)}{n} + \frac{3\zeta(-3/2)}{4n^2} + \right. \quad \text{(II.1.1)}$$

$$\left. \frac{5\zeta(-5/2)}{8n^3} + \frac{35\zeta(-7/2)}{64n^4} + \frac{63\zeta(-9/2)}{128n^5} + \frac{231\zeta(-11/2)}{512n^6} + \cdots \right)$$

$$\sum_{k=1}^{n-1} [k(n-k)]^{1/2} = \frac{n^2\pi}{8} + \sqrt{n}\left(2\zeta(-1/2) - \frac{\zeta(-3/2)}{n} - \frac{\zeta(-5/2)}{4n^2} - \right. \quad \text{(II.1.2)}$$

$$\left. \frac{\zeta(-7/2)}{8n^3} - \frac{5\zeta(-9/2)}{64n^4} - \frac{7\zeta(-11/2)}{128n^5} - \frac{21\zeta(-13/2)}{512n^6} - \cdots \right)$$

$$\sum_{k=1}^{n-1} [k(n-k)]^{-s} = (n/2)^{1-2s}\sqrt{\pi}\frac{\Gamma(1-s)}{\Gamma(3/2-s)} + 2n^{-s}\left(\zeta(s) + \right. \tag{II.1.3}$$

$$\left. \frac{s\zeta(s-1)}{n} + \frac{s(s+1)\zeta(s-2)}{2n^2} + \ldots + \frac{\Gamma(s+i)\zeta(s-i)}{\Gamma(s)i!n^i} + \ldots \right) \quad [s \neq 2,3,4,\ldots]$$

$$\sum_{k=1}^{n} k^k = n^n \left(1 + \frac{1}{en} + \frac{e+2}{2e^2n^2} + \frac{7e^2+48e+24}{24e^3n^3} + \right. \tag{II.1.4}$$

$$\frac{9e^3+160e^2+216e+48}{48e^4n^4} + \frac{743e^4+30720e^3+84240e^2+46080e+5760}{5760e^5n^5} +$$

$$\left. \frac{1075e^5+97792e^4+486000e^3+491520e^2+144000e+11520}{11520e^6n^6} + \cdots \right)$$

$$\sum_{k=1}^{n} k^n = n^n \left(\frac{e}{e-1} - \frac{e(e+1)}{2(e-1)^3 n} + \frac{e(e+5)(11e^2+2e-1)}{24(e-1)^5 n^2} + \ldots \right) \tag{II.1.5}$$

$$\sum_{k=1}^{n} k^{-s} = \zeta(s) + n^{-s}\left(\frac{n}{(1-s)} + \frac{1}{2} - \frac{s}{12n} + \frac{\Gamma(s+3)}{720\Gamma(s)n^3} - \right. \tag{II.1.6}$$

$$\left. \frac{\Gamma(s+5)}{30240\Gamma(s)n^5} - \frac{\Gamma(s+7)}{1209600\Gamma(s)n^7} \cdots \right) \quad [s \neq 1]$$

$$\sum_{k=1}^{n} \frac{z^k}{k} = -\ln(1-z) + \frac{(z-1)^{-1}z^{n+1}}{(n+1)} + \ldots + \frac{z^{n+i}(i-1)!n!}{(z-1)^i(n+i)!} + \ldots \tag{II.1.7}$$

$$= -\ln(1-z) + z^{n+1}\left(\frac{1}{(z-1)n} + \frac{1}{(z-1)^2n^2} + \frac{z+1}{(z-1)^3n^3} + \frac{z^2+4z+1}{(z-1)^4n^4} \right.$$

$$\left. + \frac{(z+1)(z^2+10z+1)}{(z-1)^5n^5} + \frac{z^4+26z^3+66z^2+26z+1}{(z-1)^6n^6} + \ldots \right) \quad [0 \leq z < 1]$$

$$\sum_{k=1}^{n} \frac{(1-z/n)^k}{k} = \ln(n/z) - E_1(z) + e^{-z}\left(\frac{z+1}{2n} - \right. \tag{II.1.8}$$

$$\frac{3z^3+z^2+2z+2}{24n^2} + \frac{z-3}{48n^3}z^4 -$$

$$\left.\frac{15z^7-135z^6+230z^5-2z^4-8z^3-24z^2-48z-48}{5760n^4}+\cdots\right) \qquad [z>0]$$

where $E_1(z)=\int_z^{\infty}\frac{e^{-t}}{t}dt$ is the exponential integral.

$$\sum_{k=1}^{\infty} z^{k^2+ak} = \frac{z^{-a^2/4}}{2}\sqrt{-\frac{\pi}{\ln z}} - \frac{a+1}{2} + a(a^2-1)\left(\frac{\ln z}{12}- \right. \qquad \text{(II.1.9)}$$

$$\frac{(a^2+1)\ln^2 z}{120}+\frac{(3a^4+3a^2+10)\ln^3 z}{5040}-\frac{(a^2+3)(a^4-2a^2+7)\ln^4 z}{30240}+$$

$$\left.\frac{(a^8+a^6+a^4-21a^2+210)\ln^5 z}{665280}+\cdots\right) + o(\ln^k z) \qquad (\textit{for any } k)$$

$$\sum_{k=1}^{\infty}\frac{1}{z^k-1} = -\log_z(z-1)+\frac{\gamma}{\ln z}+\frac{3}{4}+\frac{5\ln z}{144}-\frac{31\ln^3 z}{86400}+O(\ln^5 z) \qquad \text{(II.1.10)}$$

$$\sum_{k=1}^{\infty}\frac{z^k}{a+k} = z^{-a}\left(\psi(1+a) - \gamma - \ln(1-z) - a(z-1) - \right. \qquad \text{(II.1.11)}$$

$$\left.\frac{a(a-1)(z-1)^2}{4}-\ldots-\frac{a^{\underline{i}}\,(z-1)^i}{i\;i!}\right) \qquad [0<z<1]$$

where $a^{\underline{i}} = a(a-1)(a-2)\cdots(a-i+1)$ denotes the descending factorial.

$$\sum_{k=1}^{\infty}\frac{z^k}{k^2} = \frac{\pi^2}{6} - \ln(-\ln(z)).\ln(z) + z-1 - \frac{3\ln^2 z}{4} - \qquad \text{(II.1.12)}$$

$$\frac{13\ln^3 z}{72} - \frac{\ln^4 z}{24} - \frac{\ln^5 z}{12} - \cdots$$

II.2 Gamma-type expansions

$$\sum_{k=1}^{n}\frac{1}{k} = \psi(n+1)+\gamma = H_n = \gamma+\ln n+\frac{1}{2n}-\frac{1}{12n^2}+\frac{1}{120n^4}- \qquad \text{(II.2.1)}$$

$$\frac{1}{252n^6}+\frac{1}{240n^8}\cdots$$

$$\sum_{k=1}^{n}\frac{1}{k^2} = H_n^{(2)} = \frac{\pi^2}{6} - \frac{1}{n} + \frac{1}{2n^2} - \frac{1}{6n^3} + \frac{1}{30n^5} - \frac{1}{42n^7}\cdots \qquad \text{(II.2.2)}$$

$$\sum_{k=1}^{n}\ln k = \ln\Gamma(n+1) = (n+1/2)\ln n-n+\frac{\ln 2\pi}{2}+\frac{1}{12n}- \qquad \text{(II.2.3)}$$

$$\frac{1}{360n^3}+\frac{1}{1260n^5}-\frac{1}{1680n^7}\cdots$$

$$n! \;=\; \Gamma(n+1) \;=\; n^n\sqrt{2\pi n}\,e^{-n}\left(1+\frac{1}{12n}+\frac{1}{288n^2}-\frac{139}{51840n^3}-\frac{571}{2488320n^4}+\frac{163879}{209018880n^5}+\cdots\right) \qquad \text{(II.2.4)}$$

$$= n^n\sqrt{2\pi(n+1/6)}\,e^{-n}\left(1+\frac{1}{144n^2}+O(n^{-3})\right)$$

II.3 Exponential type expansions

$$(1+z/n)^n \;=\; e^z\left(1-\frac{z^2}{2n}+\frac{3z+8}{24n^2}z^3-\frac{(z+2)(z+6)}{48n^3}z^4+\frac{15z^3+240z^2+1040z+1152}{5760n^4}z^5-\frac{(z+4)(z^3+68z^2+408z+480)}{11520n^5}z^6+\frac{63z^5+2520z^4+35280z^3+211456z^2+526176z+414720}{2903040n^6}z^7-\cdots\right) \qquad \text{(II.3.1)}$$

$$(1+1/n)^n \;=\; e\left(1-\frac{1}{2n}+\frac{11}{24n^2}-\frac{7}{16n^3}+\frac{2447}{5760n^4}-\frac{959}{2304n^5}+\frac{238043}{580608n^6}-\cdots\right) \qquad \text{(II.3.2)}$$

$$(1-1/n)^n \;=\; e^{-1}\left(1-\frac{1}{2n}-\frac{5}{24n^2}-\frac{5}{48n^3}-\frac{337}{5760n^4}-\frac{137}{3840n^5}-\frac{67177}{2903040n^6}-\cdots\right) \qquad \text{(II.3.3)}$$

$$(1+\frac{b}{n^2})^n \;=\; 1+\frac{b}{n}+\frac{b^2}{2n^2}+\frac{b-3}{6n^3}b^2+\frac{b-12}{24n^4}b^3+\frac{b^2-30b+40}{120n^5}b^3+\frac{b^2-60b+330}{720n^6}b^4+\cdots \qquad \text{(II.3.4)}$$

$$(1+\frac{c}{n^3})^n \;=\; 1+\frac{c}{n^2}+\frac{c^2}{2n^4}-\frac{c^2}{2n^5}+\frac{c^3}{6n^6}-\frac{c^3}{2n^7}\cdots \qquad \text{(II.3.5)}$$

II.4 Asymptotic expansions of sums and definite integrals containing e^{-x^2}

$$\int_1^\infty e^{-x^2/n}dx = \frac{\sqrt{n\pi}}{2}-1+\frac{1}{3n}-\frac{1}{10n^2}+\frac{1}{42n^3}-\dots+\frac{(-1)^{i-1}}{i!(2i+1)n^i} \qquad \text{(II.4.1)}$$

$$\int_1^\infty \frac{e^{-x^2/n}}{x}dx = -\frac{\gamma}{2}+\frac{\ln n}{2}+\frac{1}{2n}-\frac{1}{8n^2}+\frac{1}{36n^3}-\dots+\frac{(-1)^{i-1}}{i!\,2i\,n^i} \qquad \text{(II.4.2)}$$

$$\int_1^\infty \frac{e^{-x^2/n}}{x^s}dx = \frac{e^{-1/n}}{s-1}-\frac{2}{(s-1)n}\int_1^\infty \frac{e^{-x^2/n}}{x^{s-2}}dx \quad (s>1) \qquad \text{(II.4.3)}$$

$$\int_0^\infty e^{-x^2/n}\ln(1+x)dx = \sqrt{n\pi}\frac{\ln(n/4)-\gamma}{4}-\frac{\gamma}{2}+1+\frac{\ln n}{2}+\frac{\sqrt{\pi/n}}{2}-\frac{\ln n+5/3-\gamma}{6n}-\frac{\sqrt{\pi/n^3}}{6} \qquad \text{(II.4.4)}$$

$$\int_0^\infty e^{-x^2/n}\ln(1+x)x\,dx = \frac{n}{2}\int_0^\infty \frac{e^{-x^2/n}}{1+x}dx \quad [c.f. II. 4.7] \qquad \text{(II.4.5)}$$

$$\int_0^\infty e^{-x^2/n}dx = \frac{\sqrt{\pi n}}{2} \qquad \text{(II.4.6)}$$

$$\int_0^\infty \frac{e^{-x^2/n}}{1+x}dx = \frac{\ln n-\gamma}{2}+\sqrt{\pi/n}-\frac{\ln n+1-\gamma}{2n}-\frac{2\sqrt{\pi/n^3}}{3}+\frac{\ln n+3/2-\gamma}{4n^2}+\frac{4\sqrt{\pi/n^5}}{15}-\dots \qquad \text{(II.4.7)}$$

$$\int_0^\infty \frac{e^{-x^2/n}}{[1+x]^s}dx = T(s) = \frac{1}{s-1}+\frac{2[T(s-2)-T(s-1)]}{n(1-s)} \quad [s>1] \qquad \text{(II.4.8)}$$

$$\sum_{k=1}^\infty \frac{e^{-k^t/n}}{k^s} = \frac{n^{\frac{1-s}{t}}\Gamma(\frac{1-s}{t})}{t}+\zeta(s)-\frac{\zeta(s-t)}{n}+\frac{\zeta(s-2t)}{2n^2}-\dots \qquad \text{(II.4.9)}$$

$[s-mt \neq 1 \text{ for } m=1,2,\cdots]$

$$= \frac{(-n)^{-m}}{tm!}(\ln n+\psi(m+1))+\zeta(s)-\frac{\zeta(s-t)}{n}+\frac{\zeta(s-2t)}{2n^2}-\dots$$

$[s-mt=1,$ and $\zeta(s-mt)$ interpreted as $\gamma]$

$$\sum_{k=1}^{\infty} e^{-k^2/n} \ln(k+1) = \sqrt{\pi n}\frac{\ln(n/4)-\gamma}{4} - \frac{\gamma}{2} + \frac{\ln 2\pi n}{2} + \frac{\sqrt{\pi/n}}{2} + \frac{G-\ln n}{6n} - \frac{\sqrt{\pi/n^3}}{6} + \cdots \qquad (II.4.10)$$

where $G=2-3\ln 2\pi-12\zeta'(-1)+6\zeta'(-2)+\gamma=-1.134052552271\cdots$

II.5 Doubly exponential forms

In the functions below, it is assumed that $P(x)$ is a periodic function with period 1.

$$\sum_{k\geq 0}\left(1-\frac{z}{n}\right)^{2^k} = -\log_2\left(\log_2\frac{n-z}{n}\right)+\frac{1}{2}+\frac{\gamma}{\ln 2}+ P\left(\log_2(\log_2\frac{n-z}{n})\right)+\frac{z}{n}+\frac{z^2}{3n^2}+\frac{4z^3}{21n^3}+\frac{41z^4}{315n^4}+\frac{136z^5}{1395n^5}+\ldots \qquad (II.5.1)$$

$$|P(x)| \leq 0.00000316\ldots$$

$$\sum_{k\geq 0}\left(1-\frac{z}{n}\right)^{\beta^k} = -\log_\beta\left(\log_\beta\frac{n-z}{n}\right)+\frac{1}{2}+\frac{\gamma}{\ln \beta}+ P\left(\log_\beta(\log_\beta\frac{n-z}{n})\right)+\frac{z}{(\beta-1)n}+\frac{\beta z^2}{2(\beta^2-1)n^2}+\frac{\beta(2+\beta+2\beta^2)z^3}{6(\beta+1)(\beta^3-1)n^3}+ \frac{\beta(6+\beta+13\beta^2+\beta^3+6\beta^4)z^4}{24(\beta^4-1)(\beta^2+\beta+1)n^4}+\ldots \qquad (II.5.2)$$

$$\sum_{k\geq 0}\left(1-e^{-ne^{-k}}\right) = \ln n+\gamma+\frac{1}{2}+P(\ln n)+e^{-n}+e^{-ne}+e^{-ne^2}+\ldots \qquad (II.5.3)$$

$$|P(x)| \leq 0.0001035$$

$$\sum_{k\geq 0}\left(1-n^{-\beta^k}\right) = -\frac{\ln(\ln n)+\gamma}{\ln \beta}+P\left(\log_\beta(\ln n)\right)+n^{-\beta^{-1}}+ n^{-\beta^{-2}}+\ldots \quad [\beta<1] \qquad (II.5.4)$$

$$\sum_{k\geq 0}\left(2^k e^{-n2^{-k}}-2^k+n\right) = n\log_2 n+1+\left(\frac{\gamma-1}{\ln 2}+\frac{1}{2}\right)n+ nP(\log_2 n)-\frac{e^{-2n}}{2}-\frac{e^{-4n}}{4}-\ldots \qquad (II.5.5)$$

$$|P(x)| \leq 0.000000173$$

II.6 Roots of polynomials

For the following polynomials we find an asymptotic expression of the real root closest to 1. We use the transcendental function $w(x)$ defined by

$$w(x)e^{w(x)} = x$$

It is known that $w(x) = \ln x - \ln(\ln x) + o(1)$ when $x \to \infty$.

$$ax^n + bx^{n-1} + f(n) = 0 \tag{II.6.1}$$

$$x = 1 + \frac{y}{n} + \frac{y}{n^2}\left(\frac{b}{a+b} + \frac{y}{2}\right) + \frac{y}{n^3}\left(\frac{b^2 + b(b+a/2)y}{(a+b)^2} + \frac{y^2}{6}\right) + O(y^4 n^{-4})$$

$$y = \ln\left(-\frac{f(n)}{a+b}\right)$$

$$(a+n)x^n + (b-n)x^{n-1} + f(n) = 0 \tag{II.6.2}$$

$$x = 1 + \frac{y-a-b}{n} + \frac{2a(a+b) + ((a+b)^2 - 4a - 2b)y + 2(1-a-b)y^2 + y^3}{2(y+1)n^2} + O(y^3 n^{-3})$$

$$y = w(-e^{a+b} f(n))$$

$$(a+n)x^n + (b-cn)x^{n-1} + f(n) = 0 \qquad [c \neq 1] \tag{II.6.3}$$

$$x = 1 + \frac{y}{n} + \frac{y}{n^2}\left(\frac{c}{c-1} + \frac{y}{2}\right) + \frac{y}{n^3}\left(\frac{b + ac + c^2 + c(c-1/2)y}{(c-1)^2} + \frac{y^2}{6}\right) + O(y^4 n^{-4})$$

$$y = \ln\left(\frac{f(n)}{(c-1)n - b - a}\right)$$

II.7 Sums containing descending factorials

For the following formulas we will denote

$$g(x) = \sum_{i \geq 0} f(x^i)$$

or

$$g(x) = \sum_{i \geq 0} a_i x^i$$

the sum being convergent in some region around 0, and

$$g_k(x) = \sum_{i \geq 0} i^{\underline{k}} f(x^i)$$

Descending factorials are denoted by $i^{\underline{k}} = i(i-1)(i-2)...(i-k+1)$.

In all cases, $\alpha = n/m$.

$$\sum_{i \ge 0} f\left(\frac{n^{\underline{i}}}{m^i}\right) = g(\alpha) + \frac{g'(\alpha) - g_1'(\alpha)}{2m} + \tag{II.7.1}$$

$$+ \frac{3\alpha(g_2''(\alpha) - g_1''(\alpha) + g''(\alpha)) - g_2'(\alpha) - g_1'(\alpha) + g'(\alpha)}{24\ mn} + O(n^{-3})$$

$$\sum_{i \ge 0} a_i \frac{n^{\underline{i}}}{m^i} = g(\alpha) - \frac{\alpha g''(\alpha)}{2m} + \frac{\alpha}{24m^2}\left(3\,\alpha\, g^{iv}(\alpha) + 8g'''(\alpha)\right) \tag{II.7.2}$$

$$- \frac{\alpha}{48m^3}\left(\alpha^2 g^{vi}(\alpha) + 8\alpha\, g^{v}(\alpha) + 12 g^{iv}(\alpha)\right) + O(m^{-4})$$

$$\sum_{i \ge 0} f\left(\frac{n^{\underline{i}}}{m^{\underline{i}}}\right) = g(\alpha) + \frac{(g'(\alpha) - g_1'(\alpha))(1-\alpha)}{2m} \tag{II.7.3}$$

$$+ \frac{(1-\alpha)[3\alpha(1-\alpha)(g_2'' - g_1'' + g'') - (7\alpha+1)g_2' + (5\alpha-1)(g_1' - g')]}{24mn} + O(n^{-3})$$

$$\sum_{i \ge 0} a_i \frac{n^{\underline{i}}}{m^{\underline{i}}} = g(\alpha) - \frac{\alpha(1-\alpha)g''(\alpha)}{2m} + \tag{II.7.4}$$

$$\frac{\alpha(1-\alpha)}{24m^2}\left[3\alpha(1-\alpha)g^{iv} + 8(1-2\alpha)g''' - 12g''\right] +$$

$$\frac{\alpha(1-\alpha)}{48m^3}\left[-\alpha^2(1-\alpha)^2 g^{vi} - 8\alpha(1-\alpha)(1-2\alpha)g^{v} - \right.$$

$$\left. 12(1-6\alpha+6\alpha^2)g^{iv} + 48(1-2\alpha)g''' - 24g''\right] + O(m^{-4})$$

$$\sum_{n} f(n^{\underline{k}}) = \left(n - \frac{k-1}{2} + \frac{1-k^2}{24n} - \frac{(k-1)^2(k+1)}{48n^2} - \right. \tag{II.7.5}$$

$$\left. \frac{(k-1)(k+1)(73k^2 - 240k + 143)}{5760n^3} + \cdots \right) T_{-1}(n^{\underline{k}}) -$$

$$\frac{f(n^{\underline{k}})}{2} + k\left(\frac{1}{12n} + \frac{k-1}{24n^2} + \frac{19(k-1)(k-2)}{720n^3} + \cdots\right) n^{\underline{k}} f'(n^{\underline{k}}) +$$

$$\left(\frac{1-k^2}{24n} - \frac{(k-1)^2(k+1)}{48n^2} - \frac{(k-1)^2(k+1)(7k-17)}{576n^3} + \cdots\right) T_1(n^{\underline{k}}) -$$

$$\frac{k^2(k-1)(n^{\underline{k}})^2 f''(n^{\underline{k}})}{240n^3} - \frac{k^3(n^{\underline{k}})^3 f'''(n^{\underline{k}})}{720n^3} -$$

$$\frac{(k-3)(k-1)(k+1)(k+3)}{640n^3} T_3(n^{\underline{k}}) + \cdots$$

where

$$T_i(x) = \frac{x^{i/k}}{k} \int f(x)\, x^{-i/k-1}\, dx$$

II.8 Summation formulas

Euler-Maclaurin summation formula

$$\sum_{k=1}^{n-1} f(k) = \int_1^n f(x)dx + \left(\sum_{i=1}^{\infty} \frac{B_i f^{(i-1)}(x)}{i!}\right)\Bigg|_{x=1}^{x=n} \tag{II.8.1}$$

where B_i are the Bernoulli numbers $B_0 = 1$, $B_1 = -1/2$, $B_2 = 1/6$, $B_4 = -1/30$, $B_6 = 1/42$, $B_8 = -1/30$, ...

$$= \int_1^n f(x)dx + \left(-\frac{f(x)}{2} + f'\frac{(x)}{12} - f'''\frac{(x)}{720} + \frac{f^{(5)}(x)}{30240} - \frac{f^{(7)}(x)}{1209600} + \cdots\right)\Bigg|_{x=1}^{x=n}$$

If we write

$$\sum_{k=1}^{n-1} f(k) = \int_a^n f(x)dx + C(f,a) + \sum_{i=1}^{\infty} \frac{B_i f^{(i-1)}(n)}{i!}$$

then, if $f(x) = \sum_i a_i x^i + \sum_i b_i \ln x\, x^i + \sum_i c_i \ln^2(x) x^i + \cdots$ (i varying over the reals),

$$C(f,1) = \sum_i a_i \phi(i) + \sum_i b_i \phi'(i) + \sum_i c_i \phi''(i) \cdots$$

$$= \left(\frac{\pi^2}{6} - 1\right) a_{-2} + \gamma a_{-1} + \frac{a_0}{2} + \frac{5a_1}{12} + \frac{a_2}{3} + \frac{31a_3}{120} + \frac{a_4}{5} + \frac{41a_5}{252} + \cdots + \frac{b_0(\ln(2\pi) - 2)}{2} + \ldots$$

where $\phi(i) = \zeta(-i) + \frac{1}{i+1}$, $\phi(-1) = \gamma$; $\phi'(-1) = -\gamma_1$, and if $a_{-1} = 0$,

$$C(f,0) = \sum_i a_i \zeta(-i) + \sum_i b_i \zeta'(-i) + \cdots$$

$$= -\frac{a_0}{2} - \frac{a_1}{12} + \frac{a_3}{120} - \frac{a_5}{252} + \cdots + \frac{b_0 \ln 2\pi}{2} \cdots$$

Appendix III: REFERENCES

III.1 Textbooks

The following are fine textbooks recommended for further information on the indicated topics.

[1] Aho, A.V., Hopcroft, J.E. and Ullman, J.D.: *The Design and Analysis of Computer Algorithms*; Addison-Wesley, Reading Mass, (1974).

[2] Aho, A.V., Hopcroft, J.E. and Ullman, J.D.: *Data Structures and Algorithms*; Addison-Wesley, Reading Mass, (1983).

[3] Baron, R.J. and Shapiro, L.G.: *Data Structures and Their Implementation*; Van Nostrand Reinhold, New York NY, (1980).

[4] Borodin, A. and Munro, J.I.: *The Computational Complexity of Algebraic and Numeric Problems*; American Elsevier, New York NY, (1975).

[5] Flores, I.: *Computer Sorting*; Prentice-Hall, Englewood Cliffs NJ, (1969).

[6] Gotlieb, C.C. and Gotlieb, L.R.: *Data Types and Structures;* Prentice-Hall, Englewood Cliffs NJ, (1978).

[7] Horowitz, E. and Sahni, S.: *Fundamentals of Data Structures;* Computer Science Press, Potomac Maryland, (1976).

[8] Johnson, N.L. and Kotz, S.: *Discrete Distributions*; Houghton Mifflin. Boston Mass., (1969).

[9] Knuth, D.E.: *The Art of Computer Programming, vol. I: Fundamental Algorithms;* Addison-Wesley, Reading Mass. (1973).(2nd edition)

[10] Knuth, D.E.: *The Art of Computer Programming, vol. II: Seminumerical Algorithms;* Addison-Wesley, Reading Mass. (1969).

[11] Knuth, D.E.: *The Art of Computer Programming, vol. III: Sorting and Searching;* Addison-Wesley, Reading Mass., (1973).

[12] Lorin, H.: *Sorting and Sort Systems*; Addison-Wesley, Reading MA, (1975).

[13] Reingold, E.M. and Hansen, W.J.: *Data Structures;* Little Brown, Boston Mass.

[14] Sedgewick, R.: *Algorithms*; Addison-Wesley, Reading Mass,(1983).

[15] Standish, T.A.: *Data Structure Techniques;* Addison-Wesley, Reading Mass. (1980).

[16] Wirth, N.: *Algorithms + Data Structures = Programs;* Prentice-Hall, Englewood Cliffs NJ, (1976).

[17] Zipf, G.K.: *Human Behaviour and the Principle of Least Effort;* Cambridge MA. Addison-Wesley, (1949).

III.2 Papers

The following are research papers that contain in-depth information on the topics covered in the indicated sections of this handbook.

[1] Ackerman, A.F.: Quadratic Search for Hash Tables of Size p^n; C.ACM, 17(3):164, (Mar 1974). (3.3.5)

[2] Adel'son-Vel'skii, G.M. and Landis, E.M.: An Algorithm for the organization of information; Dokladi Akademia Nauk SSSR, 146(2):263-266, (1962). (3.4.1.3)

[3] Adleman, L., Booth, K.S., Preparata, F.P. and Ruzzo, W.L.: Improved Time and Space Bounds for Boolean Matrix Multiplication; Acta Informatica, 11(1):61-70, (1978). (6.3)

[4] Aho, A.V., Steiglitz, K. and Ullman, J.D.: Evaluating Polynomials at Fixed Points; SIAM J on Computing, 4(4):533-539, (Dec 1975). (6.4)

[5] Ajtai, M., Fredman, M. and Komlos, J.: Hash Functions for Priority Queues; Proceedings FOCS, Tucson AZ, 24:299-303, (Nov 1983). (5.1)

[6] Ajtai, M., Komlos, J. and Szemeredi, E.: There is no Fast Single Hashing Algorithm; Inf. Proc. Letters, 7(6):270-273, (Oct 1978). (3.3.1)

[7] Akl, S.G. and Meijer, H.: On the Average-Case Complexity of Bucketing Algorithms; J. of Algorithms, 3(1):9-13, (Mar 1982). (4.2.3)

[8] Allen, B. and Munro, J.I.: Self-Organizing Search Trees; J.ACM, 25(4):526-535, (Oct 1978); also Proceedings FOCS, Houston TX, 17:166-172, (Oct 1976). (3.4.1.6, 3.1)

[9] Allen, B.: On the Costs of Optimal and Near-Optimal Binary Search Trees; Acta Informatica, 18(3):255-263, (1982). (3.4.1.6, 3.4.1.7)

[10] Allison, D.C.S. and Noga, M.T.: Selection by Distributive Partitioning; Inf. Proc. Letters, 11(1):7-8, (Aug 1980). (5.2)

[11] Allison, D.C.S. and Noga, M.T.: Usort: An Efficient Hybrid of Distributive Partitioning Sorting; BIT, 22(2):135-139, (1982). (4.2.5.2)

[12] Alt, H.: Functions Equivalent to Integer Multiplication; Proceedings Icalp, 7:30-37, (1980). (6.1)

[13] Alt, H.: Multiplication is the Easiest Nontrivial Arithmetic Function; Proceedings FOCS, Tucson AZ, 24:320-322, (Nov 1983). (6.1)

[14] Amble, O. and Knuth, D.E.: Ordered Hash Tables; Computer Journal, 17(2):135-142, (May 1974). (3.3.6)

[15] Anderson, M.R. and Anderson, M.G.: Comments on Perfect Hashing Functions: A Single Probe Retrieving Method for Static Sets; C.ACM, 22(2):104-105, (Feb 1979). (3.3)

[16] Apers, P.M.: Recursive Samplesort; BIT, 18(2):125-132, (1978). (4.1.3)

[17] Artzy, E., Hinds, J.A. and Saal, H.J.: A Fast Technique for Constant Divisors; C.ACM, 19(2):98-101, (Feb 1976). (6)

[18] Aviad, Z. and Shamir, E.: A Direct Dynamic Solution to Range Search and Related Problems for Product Regions; Proceedings FOCS, Nashville TN, 22:123-126, (Oct 1981). (3)

[19] Baer, J.L. and Schwab, B.: A Comparison of Tree-Balancing Algorithms; C.ACM, 20(5):322-330, (May 1977). (3.4.1.3, 3.4.1.4, 3.4.1.6)

[20] Baer, J.L.: Weight-Balanced Trees; Proceedings ACM-NCC, 44:467-472 (1975). (3.4.1.5)

[21] Bagchi, A. and Roy, J.K.: On V-Optimal Trees; SIAM J on Computing, 8(4):524-541, (Nov 1979). (3.4.6)

[22] Bandyopadhyay, S.K.: Comment on Weighted Increment Linear Search for Scatter Tables; C.ACM, 20(4):262-263, (Apr 1977). (3.3.3)

[23] Banerjee, J. and Ramaraman, V.: A Dual Link Data Structure for Random File Organization; Inf. Proc. Letters, 4(3):64-69, (Dec 1975). (3.3.12)

[24] Barnett, J.K.R.: A Technique for Reducing Comparison Times in Certain Applications of the Merging Method of Sorting; Inf. Proc. Letters, 2(5):127-128, (Dec 1973). (4.4)

[25] Barstow, D.R.: Remarks on A Synthesis of Several Sorting Algorithms; Acta Informatica, 13(3):225-227, (1980). (2.2.2)

[26] Batagelj, V.: The Quadratic Hash Method When the Table Size is Not a Prime Number; C.ACM, 18(4):216-217, (Apr 1975). (3.3.5)

[27] Batory, D.S.: B+Trees and Indexed Sequential Files: A Performance Comparison; Proceedings ACM-SIGMOD, :30-39, (Apr 1981). (3.4.3, 3.4.2)

[28] Bayer, R. and McCreight, E.: Organization and Maintenance of Large Ordered Indexes; Acta Informatica, 1(3):173-189, (1972). (3.4.2)

[29] Bayer, R. and Unterauer, K.: Prefix B-trees; ACM TODS, 2(1):11-26, (Mar 1977). (3.4.2)

[30] Bayer, R.: Binary B-trees for virtual memory; Proceedings ACM-SIGFIDET Workshop, San Diego CA, :219-235, (1971). (3.4.2)

[31] Bayer, R.: Symmetric Binary B-trees: Data Structure and Maintenance Algorithms; Acta Informatica, 1(4):290-306, (1972). (3.4.2.2)

[32] Bays, C.: A Note on When to Chain Overflow Items Within a Direct-Access Table; C.ACM, 16(1):46-47, (Jan 1973). (3.3.11)

[33] Bays, C.: Some Techniques for Structuring Chained Hash Tables; Computer Journal, 16(2):126-131, (May 1973). (3.3.12)

[34] Bays, C.: The Reallocation of Hash-Coded Tables; C.ACM, 16(1):11-14, (Jan 1973). (3.3)

[35] Belaga, E.G.: Some Problems Involved in the Computation of Polynomials; Dokladi Akademia Nauk SSSR, 123:775-777, (1958). (6.4)

[36] Bell, J.R. and Kaman, C.H.: The Linear Quotient Hash Code; C.ACM, 13(11):675-677, (Nov 1970). (3.3.4)

[37] Bell, J.R.: The Quadratic Quotient Method: A Hash Code Eliminating Secondary Clustering; C.ACM, 13(2):107-109, (Feb 1970). (3.3.5)

[38] Bent, S.W., Sleator, D.D. and Tarjan, R.E.: Biased 2-3 Trees; Proceedings FOCS, Syracuse NY, 21:248-254, (Oct 1980). (3.4.2.1)

[39] Bentley, J.L. and Brown, D.J.: A General Class of Resource Tradeoffs; Proceedings FOCS, Syracuse NY, 21:217-228 (Oct 1980). (2)

[40] Bentley, J.L. and Friedman, J.H.: Data Structures for Range Searching; ACM C. Surveys, 11(4):397-409, (Dec 1979). (3.4, 3.5)

[41] Bentley, J.L. and Maurer, H.A.: A Note on Euclidean Near Neighbor Searching in the Plane; Inf. Proc. Letters, 8(3):133-136, (Mar 1979). (3.5)

[42] Bentley, J.L. and Maurer, H.A.: Efficient Worst-Case Data Structures for Range Searching; Acta Informatica, 13(2):155-168, (1980). (3.5)

[43] Bentley, J.L. and Saxe, J.B.: Decomposable Searching Problems. I. Static-to-Dynamic Transformation; J. of Algorithms, 1(4):301-358, (Dec 1980). (2.2)

[44] Bentley, J.L. and Saxe, J.B.: Generating Sorted Lists of Random Numbers; ACM TOMS, 6(3):359-364, (Sep 1980). (4.2)

[45] Bentley, J.L. and Shamos, M.I.: Divide and Conquer for Linear Expected Time; Inf. Proc. Letters, 7(2):87-91, (Feb 1978). (2.2.2.1)

[46] Bentley, J.L. and Shamos, M.I.: Divide and Conquer in Multidimensional Space; Proceedings SIGACT, Hershey PA, 8:220-230, (May 1976). (2.2.2.1)

[47] Bentley, J.L. and Stanat, D.F.: Analysis of Range Searches in Quad Trees; Inf. Proc. Letters, 3(6):170-173, (July 1975). (3.5.1)

[48] Bentley, J.L. and Yao, A.C-C.: An Almost Optimal Algorithm for Unbounded Searching; Inf. Proc. Letters, 5(3):82-87, (Aug 1976). (3.2.1)

[49] Bentley, J.L.: An Introduction to Algorithm Design; IEEE Computer, :66-78, (Feb 1979). (2.2)

[50] Bentley, J.L.: Decomposable Searching Problems; Inf. Proc. Letters, 8(5):244-251, (June 1979). (2.2, 3)

[51] Bentley, J.L.: Multidimensional Binary Search Trees in Database Applications; IEEE Trans. Software Engineering, 5(4):333-340, (July 1979). (3.5.2)

[52] Bentley, J.L.: Multidimensional Binary Search Trees Used for Associative Searching; C.ACM, 18(9):509-517, (Sep 1975). (3.5.2)

[53] Bentley, J.L.: Multidimensional Divide-and-Conquer; C.ACM, 23(4):214-229, (Apr 1980). (3.5)

[54] Berman, G. and Colijn, A.W.: A Modified List Technique Allowing Binary Search; J.ACM, 21(2):227-232, (Apr 1974). (3.1.1, 3.2.1)

[55] Bini, D., Capovani, M., Romani, F. and Lotti, G.: O(n**2.7799) Complexity for n x n Approximate Matrix Multiplication; Inf. Proc. Letters, 8(5):234-235, (June 1979). (6.3)

[56] Bitner, J.R. and Huang, S.H.: Key Comparison Optimal 2-3 Trees with Maximum Utilization; SIAM J on Computing, 10(3):558-570, (Aug 1981). (3.4.2.1)

[57] Bitner, J.R.: Heuristics that dynamically organize data structures; SIAM J on Computing, 8(1):82-110, (Feb 1979). (3.1.2, 3.1.3)

[58] Blake, I.F. and Konheim, A.G.: Big Buckets Are (Are Not) Better!; J.ACM, 24(4):591-606,

(Oct 1977). (3.3.3)

[59] Blum, N., Floyd, R.W., Pratt, V., Rivest, R.L. and Tarjan, R.E.: Time Bounds for Selection; JCSS, 7(4):448-461, (Aug 1973). (5.2)

[60] Bobrow, D.G. and Clark, D.W.: Compact Encodings of List Structure; ACM TOPLAS, 1(2):266-286, (Oct 1979). (2.1)

[61] Bobrow, D.G.: A Note on Hash Linking; C.ACM, 18(7):413-415, (July 1975). (3.3)

[62] Bolour, A.: Optimal Retrieval Algorithms for Small Region Queries; SIAM J on Computing, 10(4):721-741, (Nov 1981). (3.3)

[63] Bolour, A.: Optimality Properties of Multiple-Key Hashing Functions; J.ACM, 26(2):196-210, (Apr 1979). (3.3, 3.5)

[64] Boothroyd, J.: Algorithm 201, Shellsort; C.ACM, 6(8):445, (Aug 1963). (4.1.4)

[65] Boothroyd, J.: Algorithm 207, Stringsort; C.ACM, 6(10):615, (Oct 1963). (4.1)

[66] Borodin, A. and Cook, S.: A Time-Space Tradeoff for Sorting on a General Sequential Model of Computation; Proceedings SIGACT, Los Angeles CA, 12:294-301, (Apr 1980); also SIAM J of Computing, 11(2):287-297, (May 1982). (4.1, 4.2)

[67] Borodin, A. and Cook, S.: On the Number of Additions to Compute Specific Polynomials; Proceedings SIGACT, Seattle WA, 6:342-347, (Apr 1974); also SIAM J on Computing, 5(1):146-157, (Mar 1976). (6.4)

[68] Borodin, A., Fischer, M.J., Kirkpatrick, D.G., Lynch, N.A. and Tompa, M.P.: A Time-Space Tradeoff for Sorting on Non-Oblivious Machines; Proceedings FOCS, San Juan PR, 20:319-327, (Oct 1979). (4.1, 4.2)

[69] Brent, R.P.: Multiple-Precision Zero-Finding Methods and the Complexity of Elementary Function Evaluation; in *Analytic Computational Complexity*, ed. by Traub, J.F., Academic Press, 151-176, (1976). (6.1, 6.2)

[70] Brent, R.P.: Reducing the Retrieval Time of Scatter Storage Techniques; C.ACM, 16(2):105-109, (Feb 1973). (3.3.7)

[71] Brinck, K. and Foo, N.Y.: Analysis of Algorithms on Threaded Trees; Computer Journal, 24(2):148-155, (May 1981). (3.4.1.1)

[72] Brockett, R.W. and Dobkin, D.: On the Number of Multiplications Required for Matrix Multiplication; SIAM J on Computing, 5(4):624-628, (Dec 1976). (6.3)

[73] Bron, C.: Algorithm 426: Merge Sort Algorithm (M1); C.ACM, 15(5):357-358, (May 1972). (4.2.1)

[74] Brown, M.R. and Dobkin, D.: An Improved Lower Bound on Polynomial Multiplication; IEEE Trans. on Computers, 29(5):337-340, (May 1980). (6.4)

[75] Brown, M.R. and Tarjan, R.E.: A Fast Merging Algorithm; J.ACM, 26(2):211-226, (Apr 1979). (3.4, 5.1)

[76] Brown, M.R. and Tarjan, R.E.: A Representation for Linear Lists with Movable Fingers; Proceedings SIGACT, San Diego CA, 10:19-29, (May 1978). (3.4.2.1)

[77] Brown, M.R. and Tarjan, R.E.: Design and Analysis of a Data Structure for Representing

Sorted Lists; SIAM J on Computing, 9(3):594-614, (Aug 1980). (3.4.2.1)

[78] Brown, M.R.: A Partial Analysis of Random Height-Balanced Trees; SIAM J on Computing, 8(1):33-41, (Feb 1979). (3.4.1.3)

[79] Brown, M.R.: A Storage Scheme for Height-Balanced Trees; Inf. Proc. Letters, 7(5):231-232, (Aug 1978). (3.4.1.3)

[80] Brown, M.R.: Addendum to "A Storage Scheme for Height-Balanced Trees"; Inf. Proc. Letters, 8(3):154-156, (Mar 1979). (3.4.1.3, 3.4.2.1)

[81] Brown, M.R.: Implementation and analysis of binomial queue algorithms; SIAM J on Computing, 7(3):298-319, (Aug 1978). (5.1.7)

[82] Brown, M.R.: Some Observations on Random 2-3 Trees; Inf. Proc. Letters, 9(2):57-59, (Aug 1979). (3.4.2.1)

[83] Brown, M.R.: The Complexity of Priority Queue Maintenance; Proceedings SIGACT, Boulder, 9:42-48, (May 1977). (5.1.7)

[84] Bruss, A.R. and Meyer, A.R.: On Time-Space Classes and their Relation to the Theory of Real Addition; Theoretical Computer Science, 11(1):59-69, (1980). (6)

[85] Burge, W.H.: An Analysis of Binary Search Trees Formed from Sequences of Nondistinct Keys; J.ACM, 23(3):451-454, (July 1976). (3.4.1)

[86] Burkhard, W.A. and Keller, R.M.: Some Approaches to Best-Match File Searching; C.ACM, 16(4):230-236, (Apr 1973). (3)

[87] Burkhard, W.A.: Associative Retrieval Trie Hash-Coding; Proceedings SIGACT, Hershey PA, 8:211-219, (May 1976). (3.4.4)

[88] Burkhard, W.A.: Full Table Quadratic Quotient Searching; Computer Journal, 18(1):161-163, (Feb 1975). (3.3.5)

[89] Burkhard, W.A.: Hashing and Trie Algorithms for Partial Match Retrieval; ACM TODS, 1(2):175-187, (June 1976). (3.4.4)

[90] Burkhard, W.A.: Interpolation-Based Index Maintenance; Proceedings ACM PODS, Atlanta GA, 2:76-89, (Mar 1983). (3.2.2, 3.3.13)

[91] Burkhard, W.A.: Partial-Match Hash Coding: Benefits of Redundancy; ACM TODS, 4(2):228-239, (June 1979). (3.3)

[92] Burton, F.W. and Lewis, G.N.: A Robust Variation of Interpolation Search; Inf. Proc. Letters, 10(4):198-201, (July 1980). (3.2.2)

[93] Burton, F.W..: Generalized Recursive Data Structures; Acta Informatica, 12(2):95-108, (1979). (2.1)

[94] Carter, J.L. and Wegman, M.N.: Universal Classes of Hash Functions; Proceedings SIGACT, Boulder CO, 9:106-112, (May 1977); also JCSS, 18(2):143-154, (Apr 1979). (3.3)

[95] Carter, J.L., Floyd, R.W., Gill, J., Markowsky, G. and Wegman, M.N.: Exact and Approximate Membership Testers; Proceedings SIGACT, San Diego CA, 10:59-65, (May 1978). (3)

[96] Chazelle, B.: An Improved Algorithm for the Fixed-Radius Neighbor Problem; Inf. Proc. Letters, 16(4):193-198, (May 1983). (3)

[97] Chazelle, B.: Filtering Search: A New Approach to Query-Answering; Proceedings FOCS, Tucson AZ, 24:122-132, (Nov 1983). (2.2.2.1)

[98] Chen, W-C. and Vitter, J.S.: Analysis of Early Standard Coalesced Hashing; SIAM J on Computing, to appear. (3.3.12)

[99] Chin, F.Y. and Fok, K.S.: Fast Sorting Algorithms on Uniform Ladders (Multiple Shift-Register Loops); IEEE Trans. on Computers, C29(7):618-631, (July 1980). (4.2)

[100] Chin, F.Y.: A Generalized Asymptotic Upper Bound on Fast Polynomial Evaluation and Interpolation; SIAM J on Computing, 5(4):682-690, (Dec 1976). (6.4)

[101] Choy, D.M. and Wong, C.K.: Bounds for Optimal $\alpha-\beta$ Binary Trees; BIT, 17(1):1-15, (1977). (3.4.1.7)

[102] Choy, D.M. and Wong, C.K.: Optimal $\alpha-\beta$ trees with Capacity Constraint; Acta Informatica, 10(3):273-296, (1978). (3.4.1.7)

[103] Christen, C.: Improving the Bounds on Optimal Merging; Proceedings FOCS, Ann Arbor MI, 19:259-266, (Oct 1978). (4.3.3)

[104] Cichelli, R.J.: Minimal Perfect Hash Functions Made Simple; C.ACM, 23(1):17-19, (Jan 1980). (3.3)

[105] Clapson, P.: Improving the Access Time for Random Access Files; C.ACM, 20(3):127-135, (Mar 1977). (3.3)

[106] Clark, D.W.: An Efficient List-Moving Algorithm Using Constant Workspace; C.ACM, 19(6):352-354, (June 1976). (3.1.1)

[107] Clark, K.L. and Darlington, J.: Algorithm Classification Through Synthesis; Computer Journal, 23(1):61-65, (Feb 1980). (2.2.2)

[108] Claybrook, B.G.: A Facility for Defining and Manipulating Generalized Data Structures; ACM TODS, 2(4):370-406, (Dec 1977). (2.1)

[109] Cohen, J. and Roth, M.: On the Implementation of Strassen's Fast Multiplication Algorithm; Acta Informatica, 6:341-355, (1976). (6.3)

[110] Cohen, J.: A Note on a Fast Algorithm for Sparse Matrix Multiplication; Inf. Proc. Letters, 16(5):247-248, (June 1983). (6.3)

[111] Cole, R. and Yap, C.K.: Geometric Retrieval Problems; Proceedings FOCS, Tucson AZ, 24:112-121, (Nov 1983). (3)

[112] Colin, A.J.T., McGettrick, A.D. and Smith, P.D.: Sorting Trains; Computer Journal, 23(3):270-273, (Aug 1980). (4.2, 4.4.3)

[113] Collins, G.E. and Musser, D.R.: Analysis of the Pope-Stein Division Algorithm; Inf. Proc. Letters, 6(5):151-155, (Oct 1977). (6.1)

[114] Comer, D. and Sethi, R.: The Complexity of Trie Index Construction; Proceedings FOCS, Houston TX, 17:197-207, (Oct 1976); also J.ACM, 24(3):428-440, (July 1977). (3.4.4)

[115] Comer, D.: A Note on Median Split Trees; ACM TOPLAS, 2(1):129-133, (Jan 1980).

(3.4.1.6)

[116] Comer, D.: Analysis of a Heuristic for Full Trie Minimization; ACM TODS, 6(3):513-537, (Sep 1981). (3.4.4)

[117] Comer, D.: Heuristics for Trie Index Minimization; ACM TODS, 4(3):383-395, (Sep 1979). (3.4.4)

[118] Comer, D.: The Ubiquitous B-tree; ACM C. Surveys, 11(2):121-137, (June 1979). also ACM C. Surveys, 11(4):412 (Dec 1979). (3.4.2)

[119] Cook, C.R. and Kim, D.J.: Best Sorting Algorithm for Nearly Sorted Lists; C.ACM, 23(11):620-624, (Nov 1980). (4.1)

[120] Cooper, D., Dicker, M.E. and Lynch, F.: Sorting of Textual Data Bases: A Variety Generation Approach to Distribution Sorting; Inf. Processing and Manag., 16:(49-56), (1980). (4.2.3)

[121] Coppersmith, D. and Winograd, S.: On the Asymptotic Complexity of Matrix Multiplication; Proceedings FOCS, Nashville TN, 22:82-90, (Oct 1981); also SIAM J on Computing, 11(3):472-492, (Aug 1982). (6.3)

[122] Coppersmith, D.: Rapid Multiplication of Rectangular Matrices; SIAM J on Computing, 11(3):467-471, (Aug 1982). (6.3)

[123] Cremers, A.B. and Hibbard, T.N.: Orthogonality of Information Structures; Acta Informatica, 9(3):243-261, (1978). (2.1)

[124] Culik, K.II., Ottmann, Th. and Wood, D.: Dense Multiway Trees; ACM TODS, 6(3):486-512, (Sep 1981). (3.4.2)

[125] Cunto, W.: Lower Bounds in Selection and Multiple Selection Problems; PhD Dissertation, University of Waterloo, to appear. (5.2)

[126] Darlington, J.: A Synthesis of Several Sorting Algorithms; Acta Informatica, 11(1):1-30, (1978). (2.2.2)

[127] Dasarathy, B. and Yang, C.: A Transformation on Ordered Trees; Computer Journal, 23(2):161-164, (Feb 1980). (3.4.1)

[128] Deutscher, R.F., Sorenson, P.G. and Tremblay, J.P.: Distribution dependent hashing functions and their characteristics; Proceedings ACM-SIGMOD, :224-236, (1975). (3.3)

[129] Devillers, R. and Louchard, G.: Hashing Techniques, a Global Approach; BIT, 19(3):302-311, (1979). (3.3)

[130] Devroye, L. and Klincsek, T.: Average Time Behavior of Distributive Sorting Algorithms; Computing, 26(1):1-7, (1981). (4.2.3)

[131] Dewar, R.B.K.: A Stable Minimum Storage Sorting Algorithm; Inf. Proc. Letters, 2(6):162-164, (Apr 1974). (4.2.1)

[132] Dhawan, A.K. and Srivastava, V.K.: On a New Division Algorithm; BIT, 17(4):481-485, (1977). (6.1)

[133] Dijkstra, E.W. and Gasteren, A.J.M.: An Introduction fo Three Algorithms for Sorting in Situ; Inf. Proc. Letters, 15(3):129-134, (Oct 1982). (4.1.2, 4.1.5)

[134] Dijkstra, E.W.: Smoothsort, an Alternative for Sorting In Situ; Science of Computer Programming, 1(3):223-233, (May 1982). (4.1.5)

[135] Dinsmore, R.J.: Longer Strings from Sorting; C.ACM, 8(1):48 (Jan 1965). (4.4.0.1)

[136] Doberkat, E.E.: Asymptotic Estimates for the Higher Moments of the Expected Behavior of Straight Insertion Sort; Inf. Proc. Letters, 14(4):179-182, (June 1982). (4.1.2)

[137] Doberkat, E.E.: Deleting the Root of a Heap; Acta Informatica, 17(3):245-265, (1982). (5.1.3)

[138] Doberkat, E.E.: Inserting a New Element in a Heap; BIT, 21(3):255-269, (1981). (5.1.3)

[139] Doberkat, E.E.: Some Observations on the Average Behavior of Heapsort; Proceedings FOCS, Syracuse NY, 21:229-237, (Oct 1980). (4.1.5)

[140] Dobkin, D. and Lipton, R.J.: Addition Chain Methods for the Evaluation of Specific Polynomials; SIAM J on Computing, 9(1):121-125, (Feb 1980). (6.4)

[141] Dobkin, D. and Lipton, R.J.: Multidimensional Searching Problems; SIAM J on Computing, 5(2):181-186, (June 1976). (3.5)

[142] Dobkin, D. and Lipton, R.J.: Some Generalizations of Binary Search; Proceedings SIGACT, Seattle WA, 6:310-316, (Apr 1974). (3.5)

[143] Dobkin, D. and Munro, J.I.: Determining the Mode; Theoretical Computer Science, 12:255-263, (1980). (5.2.3)

[144] Dobkin, D. and Munro, J.I.: Efficient Uses of the Past; Proceedings FOCS, Syracuse NY, 21:200-206, (Oct 1980). (3.4)

[145] Dobkin, D. and Munro, J.I.: Optimal Time Minimal Space Selection Algorithms; J.ACM, 28(3):454-461, (July 1981). (5.2)

[146] Dobkin, D. and vanLeeuwen, J.: The Complexity of Vector-Products; Inf. Proc. Letters, 4(6):149-154, (Mar 1976). (6.3)

[147] Dobkin, D.: On the Optimal Evaluation of a Set of N-Linear Forms; Proceedings SWAT (FOCS), Iowa City IO, 14:92-102, (Oct 1973). (6.3)

[148] Dobosiewicz, W.: An Efficient Variation of Bubble Sort; Inf. Proc. Letters, 11(1):5-6, (Aug 1980). (4.1.1)

[149] Dobosiewicz, W.: Sorting by Distributive Partitioning; Inf. Proc. Letters, 7(1):1-6, (Jan 1978). (4.2.5)

[150] Dobosiewicz, W.: The Practical Significance of D.P. Sort Revisited; Inf. Proc. Letters, 8(4):170-172, (Apr 1979). (4.2.5)

[151] Downey, P., Leong, B.L. and Sethi, R.: Computing Sequences with Addition Chains; SIAM J on Computing, 10(3):638-646, (Aug 1981). (6.2)

[152] Driscoll, J.R. and Lien, Y.E.: A Selective Traversal Algorithm for Binary Search Trees; C.ACM, 21(6):445-447, (June 1978). (3.4.1)

[153] Ducoin, F.: Tri par Adressage Direct; RAIRO, Informatique, 13(3):225-237, (1979). (4.1.6)

[154] Dudzinski, K. and Dydek, A.: On a Stable Minimum Storage Merging Algorithm; Inf. Proc. Letters, 12(1):5-8, (Feb 1981). (4.3.2)

[155] Dwyer, B.: One More Time-How to Update a Master File; C.ACM, 24(1):3-8, (Jan 1981). (2.2.2.1)

[156] Eastman, C.M. and Zemankova, M.: Partially Specified Nearest Neighbor Searches Using k-d Trees; Inf. Proc. Letters, 15(2):53-56, (Sep 1982). (3.5.2)

[157] Eastman, C.M.: Optimal Bucket Size for Nearest Neighbor Searching in k-d Trees; Inf. Proc. Letters, 12(4):165-167, (Aug 1981). (3.5.2)

[158] Eberlein, P.J.: A Note on Median Selection and Spider Production; Inf. Proc. Letters, 9(1):19-22, (July 1979). (5.2)

[159] Ecker, A.: The Period of Search for the Quadratic and Related Hash Methods; Computer Journal, 17(4):340-343, (Nov 1974). (3.3.5)

[160] Edelsbrunner, H. O'Rourke, J. and Seidel, R.: Constructing Arrangements of Lines and Hyperplanes with Applications; Proceedings FOCS, Tucson AZ, 24:83-91, (Nov 1983). (3)

[161] Ehrlich, G.: Searching and Sorting Real Numbers; J. of Algorithms, 2(1):1-12, (Mar 1981). (3.2.2, 4.1.6)

[162] Eppinger, J.L.: An Empirical Study of Insertion and Deletion in Binary Search Trees; C.ACM, 26(9):663-669, (Sep 1983). (3.4.1.1)

[163] Er, M.C. and Lowden, B.G.T.: The Theory and Practice of Constructing an Optimal Polyphase Sort; Computer Journal, 25(1):93-101, (Feb 1982). (4.4.3)

[164] Erkio, H.: A Heuristic Approximation of the Worst Case of Shellsort; BIT, 20(2):130-136, (1980). (4.1.4)

[165] Erkio, H.: Internal Merge Sorting with Delayed Selection; Inf. Proc. Letters, 11(3):137-140, (Nov 1980). (4.2.1)

[166] Erkio, H.: Speeding Sort Algorithms by Special Instructions; BIT, 21(1):2-19, (1981). (4.1)

[167] Espelid, T.O.: Analysis of a Shellsort Algorithm; BIT, 13(4):394-400, (1973). (4.1.4)

[168] Espelid, T.O.: On Replacement Selection and Dinsmore's Improvement; BIT, 16(2):133-142, (1976). (4.4.0.1)

[169] Eve, J.: The Evaluation of Polynomials; Numer Math, 6:17-21, (1974). (6.4)

[170] Fabri, J.: Some Remarks on p-Way Merging; SIAM J on Computing, 6(2):268-271, (June 1977). (4.3)

[171] Fagin, R., Nievergelt, J. Pippenger, N. and Strong, H.R.: Extendible Hashing- A Fast Access Method for Dynamic Files; ACM TODS, 4(3):315-344, (Sep 1979). (3.3.13)

[172] Feig, E.: Minimal Algorithms for Bilinear Forms May Have Divisions; J. of Algorithms, 4(1):81-84, (Mar 1983). (6.3)

[173] Feig, E.: On Systems of Bilinear Forms Whose Minimal Division-Free Algorithms are all Bilinear; J. of Algorithms, 2(3):261-281, (Sep 1981). (6.3)

[174] Feldman, J.A. and Low, J.R.: Comment on Brent's Scatter Storage Algorithm; C.ACM, 16(11):703, (Nov 1973). (3.3.7)

[175] Finkel, R.A. and Bentley, J.L.: Quad Trees A Data Structure for Retrieval on Composite Keys; Acta Informatica, 4(1):1-9, (1974). (3.5.1)

[176] Flajolet, P. and Martin, N.G.: Probabilistic Counting; Proceedings FOCS, Tucson AZ, 24:76-82, (Nov 1983). (6.1)

[177] Flajolet, P. and Odlyzko, A.: Exploring Binary Trees and Other Simple Trees; Proceedings FOCS, Syracuse NY, 21:207-216, (Oct 1980). (3.4.1.2)

[178] Flajolet, P. and Puech, C.: Tree Structures for Partial Match Retrieval; Proceedings FOCS, Tucson AZ, 24:282-288, (Nov 1983). (3.5.1, 3.5.2)

[179] Flajolet, P. and Steyaert, J.M.: A Branching Process Arising in Dynamic Hashing, Trie Searching and Polynomial Factorization; Proceedings Icalp, Aarhus, 9:239-251, (July 1982). (3.3.19, 3.4.4)

[180] Flajolet, P., Francon, J. and Vuillemin, J.: Computing Integrated Costs of Sequences of Operations with Applications to Dictionaries; Proceedings SIGACT, Atlanta GA, 11:49-61, (Apr 1979). (3.1.1, 3.2.1, 3.4.1)

[181] Flajolet, P., Francon, J. and Vuillemin, J.: Sequence of Operations Analysis for Dynamic Data Structures; J. of Algorithms, 1(2):111-141, (June 1980). (3.1.1, 3.2, 3.4.1, 5.1)

[182] Flajolet, P., Francon, J. and Vuillemin, J.: Towards Analysing Sequences of Operations for Dynamic Data Structures; Proceedings FOCS, San Juan PR, 20:183-195, (Oct 1979). (3.1.1, 3.2, 3.4.1, 5.1)

[183] Flores, I. and Madpis, G.: Average Binary Search Length for Dense Ordered Lists; C.ACM, 14(9):602-603, (Sep 1971). (3.2.1)

[184] Flores, I.: Analysis of Internal Computer Sorting; J.ACM, 8(1):41-80, (Jan 1961). (4.1)

[185] Flores, I.: Computer Time for Address Calculation Sorting; J.ACM, 7(4):389-409, (Oct 1960). (4.1.6, 4.2.3)

[186] Floyd, R.W. and Rivest, R.L.: Expected Time Bounds for Selection; C.ACM, 18(3):165-172, (Mar 1975). (5.2, 5.3)

[187] Floyd, R.W. and Smith, A.J.: A Linear Time Two Tape Merge; Inf. Proc. Letters, 2(5):123-125, (Dec 1973). (4.3)

[188] Floyd, R.W.: Algorithm 245, Treesort3; C.ACM, 7(12):701, (Dec 1964). (4.1.5, 5.1.3)

[189] Floyd, R.W.: The Exact Time Required to Perform Generalized Addition; Proceedings FOCS, Berkeley CA, 16:3-5, (Oct 1975). (6.1)

[190] Foster, C.C.: A Generalization of AVL Trees; C.ACM, 16(8):513-517, (Aug 1973). (3.4.1.3)

[191] Foster, C.C.: Information Storage and Retrieval Using AVL Trees; Proceedings ACM-NCC, :192-205, (1965). (3.4.1.3)

[192] Francon, J., Viennot G. and Vuillemin J.: Description and Analysis of an Efficient Priority Queue Representation; Proceedings FOCS, Ann Arbor MI, 19:1-7, (Oct 1978). (5.1.5)

[193] Franklin, W.R.: Padded Lists: Set Operations in Expected O(log log N) Time; Inf. Proc. Letters, 9(4):161-166, (Nov 1979). (3.2.2)

[194] Franta, W.R. and Maly, K.: A Comparison of Heaps and the TL Structure for the Simulation Event Set; C.ACM, 21(10):873-875, (Oct 1978). (5.1.9)

[195] Franta, W.R. and Maly, K.: An Efficient Data Structure for the Simulation Event Set; C.ACM, 20(8):596-602, (Aug 1977). (5.1.9)

[196] Frazer, W.D. and Bennett, B.T.: Bounds of Optimal Merge Performance, and a Strategy for Optimality; J.ACM, 19(4):641-648, (Oct 1972). (4.4)

[197] Frazer, W.D. and McKellar, A.C.: Samplesort: A Sampling Approach to Minimal Storage Tree Sorting; J.ACM, 17(3):496-507, (July 1970). (4.1.3)

[198] Frazer, W.D. and Wong, C.K.: Sorting by Natural Selection; C.ACM, 15(10):910-913, (Oct 1972). (4.4.0.2)

[199] Frederickson, G.N. and Johnson, D.B.: Generalized Selection and Ranking; Proceedings SIGACT, Los Angeles CA, 12:420-428, (Apr 1980). (5.2)

[200] Frederickson, G.N.: Implicit Data Structures for the Dictionary Problem; J.ACM, 30(1):80-94, (Jan 1983).(3)

[201] Frederickson, G.N.: Implicit Data Structures for the Weighted Dictionary Problem; Proceedings FOCS, Nashville TN, 22:133-139, (Oct 1981).(3)

[202] Frederickson, G.N.: Implicit Data Structures with Fast Update; Proceedings FOCS, Syracuse, NY, 21:255-259 (Oct 1980).(3)

[203] Fredkin, E.: Trie Memory; C.ACM, 3(9):490-499, (Sep 1960). (3.4.4)

[204] Fredman, M.L., Komlos, J. and Szemeredi, E.: Storing a Sparse Table with O(1) Worst Case Access Time; Proceedings FOCS, Chicago IL, 23:165-169, (Oct 1982). (3.3)

[205] Fredman, M.L.: A Lower Bound on the Complexity of Orthogonal Range Queries; J.ACM, 28(4):696-705, (Oct 1981). (3)

[206] Fredman, M.L.: A Near Optimal Data Structure for a Type of Range Query Problem; Proceedings SIGACT, Atlanta GA, 11:62-66, (Apr 1979). (3)

[207] Fredman, M.L.: Lower Bounds on the Complexity of Some Optimal Data Structures; SIAM J on Computing, 10(1):1-10, (Feb 1981). (3)

[208] Fredman, M.L.: The Complexity of Maintaining an Array and Computing its Partial Sums; J.ACM, 29(1):250-260, (Jan 1982). (3)

[209] Fredman, M.L.: The Inherent Complexity of Dynamic Data Structures Which Accommodate Range Queries; Proceedings FOCS, Syracuse NY, 21:191-199, (Oct 1980). (3)

[210] Fredman, M.L.: Two Applications of a Probabilistic Search Technique: Sorting X+Y and Building Balanced Search Trees; Proceedings SIGACT, Albuquerque NM, 7:240-244, (May 1975). (3.4.1.6)

[211] Friend, E.H.: Sorting on Electronic Computer Systems; J.ACM, 3(3):134-168, (July 1956). (4.1, 4.2, 4.4)

[212] Fussenegger, F. and Gabow, H.N.: A Counting Approach to Lower Bounds for Selection Problems; J.ACM, 26(2):227-238, (Apr 1979). (5.2)

[213] Fussenegger, F. and Gabow, H.N.: Using Comparison Trees to Derive Lower Bounds for Selection Problems; Proceedings FOCS, Houston TX, 17:178-182, (Oct 1976). (5.2)

[214] Gairola, B.K. and Rajaraman, V.: A Distributed Index Sequential Access Method; Inf.

Proc. Letters, 5(1):1-5, (May 1976). (3.4.3)

[215] Galil, Z. and Megiddo, N.: A Fast Selection Algorithm and the Problem of Optimum Distribution of Effort; J.ACM, 26(1):58-64, (Jan 1979). (5)

[216] Gamzon, E. and Picard, C.F: Algorithme de Tri par Adressage Direct; C.R. Academie Sc. Paris, 269(A):38-41 (July 1969). (4.1.6)

[217] Garey, M.R.: Optimal Binary Search Trees with Restricted Maximal Depth; SIAM J on Computing, 3(2):101-110, (June 1974). (3.4.1.7)

[218] Gargantini, I.: An Effective Way to Represent Quadtrees; C.ACM, 25(12):905-910, (Dec 1982). (3.5.1)

[219] Garsia, A.M. and Wachs, M.L.: A New Algorithm for Minimum Cost Binary Trees; SIAM J on Computing, 6(4):622-642, (Dec 1977). (3.4.1.7)

[220] Gassner, B.J.: Sorting by Replacement Selecting; C.ACM, 10(2):89-93, (Feb 1967). (4.4.0.1)

[221] Ghosh, S., and Senko, M.: File Organization: On the Selection of Random Access Index Points for Sequential Files; J.ACM, 16(4):569-579, (Oct 1969). (3.4.3)

[222] Ghoshdastidar, D. and Roy, M.K.: A Study on the Evaluation of Shell's Sorting Technique; Computer Journal, 18(3):234-235, (Aug 1975). (4.1.4)

[223] Gill, A.: Hierarchical Binary Search; C.ACM, 23(5):294-300, (May 1980). (3.4.1)

[224] Gilstad, R.L.: Polyphase Merge Sorting - an Advanced Technique; Proceedings AFIPS Eastern J.C.C., 18:143-148, (Dec 1960). (4.4.3)

[225] Gilstad, R.L.: Read-Backward Polyphase Sorting; C.ACM, 6(5):220-223, (May 1963). (4.4.3)

[226] Goetz, M.A. and Toth, G.S.: A Comparison Between the Polyphase and Oscillating Sort Techniques; C.ACM, 6(5):223-225, (May 1963). (4.4.4)

[227] Goetz, M.A.: Internal and Tape Sorting Using the Replacement-Selection Technique; C.ACM, 6(5):201-206, (May 1963). (4.4.0.1)

[228] Gonnet, G.H. and Larson, P.A.: External Hashing with Limited Internal Storage; Proceedings ACM PODS, Los Angeles CA, 1:256-261, (Mar 1982). (3.3.15)

[229] Gonnet, G.H. and Munro, J.I.: A Linear Probing Sort and its Analysis; Proceedings SIGACT, Milwaukee WI, 13:90-95, (May 1981). (4.1.7)

[230] Gonnet, G.H. and Munro, J.I.: Efficient Ordering of Hash Tables; SIAM J on Computing, 8(3):463-478, (Aug 1979). (3.3.9, 3.3.8)

[231] Gonnet, G.H. and Munro, J.I.: Heaps on Heaps; Proceedings Icalp, Aarhus, 9:282-291, (July 1982). (5.1.3)

[232] Gonnet, G.H. and Munro, J.I.: The Analysis of an Improved Hashing Technique; Proceedings SIGACT, Boulder CO, 9:113-121, (May 1977). (3.3.8, 3.3.9)

[233] Gonnet, G.H. and Rogers, L.D.: The Interpolation-Sequential Search Algorithm; Inf. Proc. Letters, 6(4):136-139, (Aug 1977). (3.2.3)

[234] Gonnet, G.H., Munro, J.I. and Suwanda, H.: Exegesis of Self-Organizing Linear Search;

SIAM J on Computing, 10(3):613-637, (Aug 1981). (3.1.2, 3.1.3)

[235] Gonnet, G.H., Munro, J.I. and Suwanda, H.: Toward Self-Organizing Linear Search; Proceedings FOCS, San Juan PR, 20:169-174, (Oct 1979). (3.1.2, 3.1.3)

[236] Gonnet, G.H., Olivie, H.J. and Wood, D.: Height-Ratio-Balanced Trees; Computer Journal, 26(2):106-108, (May 1983). (3.4.1.3)

[237] Gonnet, G.H., Rogers, L.D. and George, A.: An Algorithmic and Complexity Analysis of Interpolation Search; Acta Informatica, 13(1):39-46, (Jan 1980). (3.2.2)

[238] Gonnet, G.H.: Average Lower Bounds for Open Addressing Hash Coding; Proceedings Theoretical Comp. Science, Waterloo Ont, :159-162, (Aug 1977). (3.3.9)

[239] Gonnet, G.H.: Expected Length of the Longest Probe Sequence in Hash Code Searching; J.ACM, 28(2):289-304, (Apr 1981). (3.3.1, 3.3.9, 3.3.10)

[240] Gonnet, G.H.: Heaps Applied to Event Driven Mechanisms; C.ACM, 19(7):417-418, (July 1976). (5.1.3)

[241] Gonnet, G.H.: Interpolation and Interpolation-Hash Searching; PhD Dissertation, University of Waterloo, (Feb 1977). (3.2.2)

[242] Gonnet, G.H.: Notes on the Derivation of Asymptotic Expressions from Summations; Inf. Proc. Letters, 7(4):165-169, (June 1978). (9.6)

[243] Gonnet, G.H.: On Direct Addressing Sort; T.S.I, to appear. (4.1.6)

[244] Gonnet, G.H.: Open Addressing Hashing with Unequal Probability Keys; JCSS, 21(3):354-367, (Dec 1980). (3.3.1)

[245] Gonzalez, T.F. and Johnson, D.B.: Sorting Numbers in Linear Expected Time and Optimal Extra Space; Inf. Proc. Letters, 15(3):119-124, (Oct 1982), also in Proceedings 16th Allerton Conference, (1978). (4.1.8)

[246] Gotlieb, C.C. and Walker, W.A.: A Top-Down Algorithm for Constructing Nearly Optimal Lexicographical Trees; in Graph Theory and Computing, Academic Press, NY, (1972). (3.4.1.6)

[247] Gotlieb, C.C.: Sorting on Computers; C.ACM, 6(5):194-201, (May 1963). (4.4)

[248] Gotlieb, L.R.: Optimal Multi-Way Search Trees; SIAM J on Computing, 10(3):422-433, (Aug 1981). (3.4.2)

[249] Guibas, L.J. and Sedgewick, R.: A Dichromatic Framework for Balanced Trees; Proceedings FOCS, Ann Arbor MI, 19:8-21, (Oct 1978). (3.4.1.3, 3.4.2)

[250] Guibas, L.J. and Szemeredi, E.: The Analysis of Double Hashing; Proceedings SIGACT, Hershey PA, 8:187-191, (May 1976); also JCSS 16(2):226-274, (Apr 1978). (3.3.4)

[251] Guibas, L.J., McCreight, E.M., Plass, M.F., and Roberts, J.R.: A New Representation for Linear Lists; Proceedings SIGACT, Boulder CO, 9:49-60, (May 1977). (3.2, 3.4.2)

[252] Guibas, L.J.: A Principle of Independence for Binary Tree Searching; Acta Informatica, 4:293-298, (1975). (3.4.1.1)

[253] Guibas, L.J.: The Analysis of Hashing Algorithms; PhD Dissertation, Stanford University, (Aug 1976). (3.3.4, 3.3)

[254] Guibas, L.J.: The Analysis of Hashing Techniques that Exhibit k-ary Clustering; Proceedings FOCS, Houston TX, 17:183-196, (Oct 1976); also J.ACM , 25(4):544-555, (Oct 1978). (3.3)

[255] Gupta, U.I., Lee, D.T. and Wong, C.K.: Ranking and Unranking of 2-3 Trees; SIAM J on Computing, 11(3):582-590, (Aug 1982). (3.4.2.1)

[256] Gupta, U.I., Lee, D.T. and Wong, C.K.: Ranking and Unranking of B-trees; J. of Algorithms, 4(1):51-60, (Mar 1983). (3.4.2)

[257] Guting, H. and Kriegel, H.P.: Dynamic k-dimensional Multiway Search Under Time-varying Access Frequencies; Lecture Notes in Computer Science, 104:135-145, (1981). (3.5)

[258] Guting, H. and Kriegel, H.P.: Multidimensional B-tree: An Efficient Dynamic File Structure for Exact Match Queries; Informatik Fachberichte, 33:375-388, (1980). (3.5)

[259] Halatsis, C. and Philokypru, G.: Pseudo Chaining in Hash Tables; C.ACM, 21(7):554-557, (July 1978). (3.3)

[260] Hansen, W.J.: A Cost Model for the Internal Organization of B+ Tree Nodes; ACM TOPLAS, 3(4):508-532, (Oct 1981). (3.4.2)

[261] Hansen, W.J.: A Predecessor Algorithm for Ordered Lists; Inf. Proc. Letters, 7(3):137-138, (Apr 1978). (3.1.1)

[262] Harper, L.H., Payne, T.H., Savage, J.E. and Straus, E.: Sorting X+Y; C.ACM, 18(6):347-349, (June 1975). (4.2, 4.3)

[263] Head, A.K.: Multiplication Modulo n; BIT, 20(1):115-116, (1980). (6.1)

[264] Heintz, J. and Schnorr, C.P.: Testing Polynomials Which are Easy to Compute; Proceedings SIGACT, Los Angeles CA, 12:262-272, (Apr 1980). (6.4)

[265] Heintz, J. and Sieveking, M.: Lower Bounds for Polynomials with Algebraic Coefficients; Theoretical Computer Science, 11:321-330, (1980). (6.4)

[266] Heising, W.P.: Note on Random Addressing Techniques; IBM Systems J, 2(2):112-116, (June 1963). (I.4)

[267] Held, G. and Stonebraker, M.: B-trees re-examined; C.ACM, 21(2):139-143, (Feb 1978). (3.4.2)

[268] Hendricks, W.J.: An account of self-organizing systems; SIAM J on Computing, 5(4):715-723, (Dec 1976). (3.1.2, 3.1.3)

[269] Hertel, S.: Smoothsort's Behavior on Presorted Sequences; Inf. Proc. Letters, 16(4):165-170, (May 1983). (4.1.5)

[270] Hibbard, T.N.: An Empirical Study of Minimal Storage Sorting; C.ACM, 6(5):206-213, (May 1963). (4.1, 4.2.4)

[271] Hirschberg, D.S.: An Insertion Technique for One-Sided Height-Balanced Trees; C.ACM, 19(8):471-473, (Aug 1976). (3.4.1.3)

[272] Hirschberg, D.S.: Fast Parallel Sorting Algorithms; C.ACM, 21(8):657-661, (Aug 1978). (4)

[273] Hirschberg, D.S.: On the Complexity of Searching a Set of Vectors; SIAM J on Computing, 9(1):126-129, (Feb 1980). (3.5)

[274] Hoare, C.A.R.: Algorithm 63 and 64; C.ACM, 4(7):321, (July 1961). (4.1.3)

[275] Hoare, C.A.R.: Algorithm 65 (FIND); C.ACM, 4(7):321-322, (July 1961). (5.2)

[276] Hoare, C.A.R.: Quicksort; Computer Journal, 5(4):10-15, (Apr 1962). (4.1.3)

[277] Horibe, Y. and Nemetz, T.O.H.: On the Max-Entropy Rule for a Binary Search Tree; Acta Informatica, 12(1):63-72, (1979). (3.4.1.6)

[278] Horowitz, E.: A Unified View of the Complexity of Evaluation and Interpolation; Acta Informatica, 3(2):123-133, (1974). (6.4)

[279] Horowitz, E.: The Efficient Calculation of Powers of Polynomials; Proceedings FOCS, College Park Md, 13:97-104, (Oct 1972); also JCSS, 7(5):469-480, (Oct 1973). (6.2, 6.4)

[280] Horvath, E.C.: Some Efficient Stable Sorting Algorithms; Proceedings SIGACT, Seattle WA, 6:194-215, (Apr 1974). (4.3.2)

[281] Horvath, E.C.: Stable Sorting in Asymptotically Optimal Time and Extra Space; J.ACM, 25(2):177-199, (Apr 1978). (4.1, 4.3.2)

[282] Hoshi, M. and Yuba, T.: A Counter Example to a Monotonicity Property of k-d Trees; Inf. Proc. Letters, 15(4):169-173, (Oct 1982). (3.5.2)

[283] Hosken, W.H.: Optimum Partitions of Tree Addressing Structures; SIAM J on Computing, 4(3):341-347, (Sep 1975). (3.4.1.7)

[284] Hu, T.C. and Shing, M.T.: Computation of Matrix Chain Products. Part I; SIAM J on Computing, 11(2):362-373, (May 1982). (6.3)

[285] Hu, T.C. and Tan, K.C.: Least Upper Bound on the Cost of Optimum Binary Search Trees; Acta Informatica, 1(4):307-310, (1972). (3.4.1.7)

[286] Hu, T.C. and Tucker, A.C.: Optimal Computer Search Trees and Variable-Length Alphabetical Codes; SIAM J. Appl Math, 21(4):514-532, (Dec 1971). (3.4.1.7)

[287] Hu, T.C., Kleitman, D.J. and Tamaki, J.K.: Binary Trees Optimum Under Various Criteria; SIAM J. Appl Math, 37(2):246-256, (Oct 1979). (3.4.1.7)

[288] Hu, T.C.: A New Proof of the T-C Algorithm; SIAM J. Appl Math, 25(1):83-94, (July 1973). (3.4.1.7)

[289] Huddleston, S. and Mehlhorn, K.: A New Data Structure for Representing Sorted Lists; Acta Informatica, 17(2):157-184, (1982). (3.4.2)

[290] Huddleston, S. and Mehlhorn, K.: Robust Balancing in B-Trees; Lecture Notes in Computer Science, 104:234-244, (1981). (3.4.2)

[291] Huits, M. and Kumar, V.: The Practical Significance of Distributive Partitioning Sort; Inf. Proc. Letters, 8(4):168-169, (Apr 1979). (4.2.5)

[292] Hwang, F.K. and Lin, S.: A Simple Algorithm for Merging Two Disjoint Linearly Ordered Sets; SIAM J on Computing, 1(1):31-39, (Mar 1972). (4.3.3)

[293] Hwang, F.K. and Lin, S.: Optimal Merging of 2 Elements with *n* Elements; Acta Informatica, 1(2):145-158, (1971). (4.3.3)

[294] Hwang, F.K.: Optimal Merging of 3 Elements with n Elements; SIAM J on Computing, 9(2):298-320, (May 1980). (4.3.3)

[295] Hwang, K. and Yao, S.B.: Optimal Batched Searching of Tree Structured Files in Multiprocessor Computer Systems; J.ACM, 24(3):441-454, (July 1977). (3.4)

[296] Hyafil, L. and Wiele, J.P. van de: On the Additive Complexity of Specific Polynomials; Inf. Proc. Letters, 4(2):45-47, (Nov 1975). (6.4)

[297] Hyafil, L., Prusker, F. and Vuillemin, J.: An Efficient Algorithm for Computing Optimal Disk Merge Patterns; Proceedings SIGACT, Seattle WA, 6:216-229, (Apr 1974). (4.3, 4.4)

[298] Hyafil, L.: Bounds for Selection; SIAM J on Computing, 5(1):109-114, (Mar 1976). (5.2)

[299] Incerpi, J. and Sedgewick, R.: Improved Upper Bounds on Shellsort; Proceedings FOCS, Tucson AZ, 24:48-55, (Nov 1983). (4.1.4)

[300] Isaac, E.J. and Singleton, R.C.: Sorting by Address Calculation; J.ACM, 3(3):169-174, (July 1956). (4.1.6, 4.2.3)

[301] Itai, A.: Optimal Alphabetic Trees; SIAM J on Computing, 5(1):9-18, (Mar 1976). (3.4.1.7)

[302] Ja'Ja', J.: On the Complexity of Bilinear Forms with Commutativity; Proceedings SIGACT, Atlanta GA, 11:197-208, (Apr 1979); also SIAM J on Computing, 9(4):713-728, (Nov 1980). (6.3)

[303] Ja'Ja', J.: On the Computational Complexity of the Permanent; Proceedings FOCS, Tucson AZ, 24:312-319, (Nov 1983). (6.3)

[304] Ja'Ja', J.: Optimal Evaluation of Pairs of Bilinear Forms; Proceedings SIGACT, San Diego CA, 10:173-183, (May 1978); also SIAM J on Computing, 8(3):443-462, (Aug 1979). (6.1, 6.3)

[305] Jackowski, B.L., Kubiak, R. and Sokolowski, S.: Complexity of Sorting by Distributive Partitioning; Inf. Proc. Letters, 9(2):100, (Aug 1979). (4.2.5)

[306] Jaeschke, G.: Reciprocal Hashing: A Method for Generating Minimal Perfect Hashing Functions; C.ACM, 24(12):829-833, (Dec 1981). (3.3)

[307] Janko, W.: Variable Jump Search: The Algorithm and its Efficiency; Angewandte Informatik, 23(1):6-11, (Jan 1981). (3.1.5)

[308] Johnson, D.B. and Kashdan, S.D.: Lower Bounds for Selection in X+Y and Other Multisets; J.ACM, 25(4):556-570, (Oct 1978). (5)

[309] Johnson, D.B. and Mizoguchi, T.: Selecting the Kth Element in X+Y and X1+X2+...+Xm; SIAM J on Computing, 7(2):147-153, (May 1978). (5.2)

[310] Johnson, D.B.: Priority Queues with Update and Finding Minimum Spanning Trees; Inf. Proc. Letters, 4(3):53-57, (Dec 1975). (5.1)

[311] Johnson, L.R.: An Indirect Chaining Method for Addressing on Secondary Keys; C.ACM, 4(5):218-222, (May 1961). (3.3.11)

[312] Jonassen, A.T. and Dahl, O-J.: Analysis of an Algorithm for Priority Queue Administration; BIT, 15(4):409-422, (1975). (5.1.2)

[313] Jonassen, A.T. and Knuth, D.E.: A Trivial Algorithm Whose Analysis Isn't; JCSS, 16(3):301-322, (June 1978). (3.4.1.1)

[314] Jones, B.: A Variation on Sorting by Address Calculation; C.ACM, 13(2):105-107, (Feb 1970). (4.1.6, 4.2.1)

[315] Jones, P.R.: Comment on Average Binary Search Length; C.ACM, 15(8):774, (Aug 1972). (3.2.1)

[316] Kahaner, D.K.: Algorithm 561-Fortran Implementation of Heap Programs for Efficient Table Maintenance; ACM TOMS, 6(3):444-449, (Sep 1980). (5.1.3)

[317] Karlton, P.L., Fuller, S.H., Scroggs, R.E. and Kaehler, E.B.: Performance of Height-Balanced Trees; C.ACM, 19(1):23-28, (Jan 1976). (3.4.1.3)

[318] Kedem, Z.M.: Combining Dimensionality and Rate of Growth Arguments for Establishing Lower Bounds on the Number of Multiplications; Proceedings SIGACT, Seattle WA, 6:334-341, (Apr 1974). (6.2)

[319] Keehn, D.G. and Jacy, J.O.: VSAM Data Set Design Parameters; IBM Systems J, 13(3):186-212, (1974). (3.4.3)

[320] Kemp, R.: A Note on the Stack Size of Regularly Distributed Binary Trees; BIT, 20(2):157-163, (1980). (3.4.1.2)

[321] Kemp, R.: The Average Number of Registers Needed to Evaluate a Binary Tree Optimally; Acta Informatica, 11(4):363-372, (1979). (3.4.1.2)

[322] Kennedy, S.: A Note on Optimal Doubly-Chained Trees; C.ACM, 15(11):997-998, (Nov 1972). (3.4.1.7)

[323] Kirkpatrick, D.G.: A Unified Lower Bound for Selection and Set Partitioning Problems; J.ACM, 28(1):150-165, (Jan 1981). (5.2)

[324] Kleitman, D.J. and Saks, M.E.: Set Orderings Requiring Costliest Alphabetic Binary Trees; SIAM J. Alg Disc Methods, 2(2):142-146, (June 1981). (3.4.1.7)

[325] Kleitman, D.J., Meyer, A.R., Rivest, R.L., Spencer, J. and Winklmann, K.: Coping with Errors in Binary Search Procedures; Proceedings SIGACT, San Diego CA, 10:227-232, (May 1978). (3.4.1)

[326] Knott, G.D.: A Numbering System for Binary Trees; C.ACM, 20(2):113-115, (Feb 1977). (3.4.1)

[327] Knott, G.D.: Fixed-Bucket Binary Storage Trees; J. of Algorithms, 3(3):276-287, (Sep 1982). (3.4.1.1, 3.4.4)

[328] Knott, G.D.: Hashing Functions; Computer Journal, 18(3):265-278, (Aug 1975). (3.3)

[329] Knuth, D.E.: Deletions that Preserve Randomness; IEEE Trans. Software Engineering, 3:351-359, (1977). (3.4.1.1)

[330] Knuth, D.E.: Evaluating Polynomials by Computers; C.ACM, 5:595-599, (1962). (6.4)

[331] Knuth, D.E.: Length of Strings for a Merge Sort; C.ACM, 6(11):685-688, (Nov 1963). (4.4.0.1, 4.4.0.3)

[332] Knuth, D.E.: Optimum Binary Search Trees; Acta Informatica, 1(1):14-25, (1971).

(3.4.1.7)

[333] Knuth, D.E.: Structured Programming with Go To Statements; ACM C. Surveys, 6(4):261-301, (Dec 1974). (3.1.1, 3.4.1.1, 4.1, 4.1.3)

[334] Knuth, D.E.: The Average Time for Carry Propagation; P. Kon Ned A, 81(2):238-242, (1978). (6.1)

[335] Konheim, A.G. and Weiss, B.: An Occupancy Discipline and Applications; SIAM J. Appl Math, 14:1266-1274, (1966). (3.3.3)

[336] Korsh, J.F.: Greedy Binary Search Trees are Nearly Optimal; Inf. Proc. Letters, 13(1):16-19, (Oct 1981). (3.4.1.6)

[337] Korsh, J.F.: Growing Nearly Optimal Binary Search Trees; Inf. Proc. Letters, 14(3):139-143, (May 1982). (3.4.1.6)

[338] Kosaraju, S.R.: Insertions and Deletions in One-Sided Height-Balanced Trees; C.ACM, 21(3):226-227, (Mar 1978). (3.4.1.3)

[339] Kosaraju, S.R.: Localized Search in Sorted Lists; Proceedings SIGACT, Milwaukee WI, 13:62-69, (May 1981). (3.4.2.1)

[340] Kosaraju, S.R.: On a Multidimensional Search Problem; Proceedings SIGACT, Atlanta GA, 11:67-73, (Apr 1979). (3.5.3)

[341] Kriegel, H.P. and Kwong, Y.S.: Insertion-Safeness in Balanced Trees; Inf. Proc. Letters, 16(5):259-264, (June 1983). (3.4.2.1)

[342] Kriegel, H.P., Vaishnavi, V.K. and Wood, D.: 2-3 Brother Trees; BIT, 18(4):425-435, (1978). (3.4.2.1)

[343] Kritzinger, P.S. and Graham, J.W.: A Theorem in the Theory of Compromise Merge Methods; J.ACM, 21(1):157-160, (Jan 1974). (4.4.3, 4.4.2)

[344] Kronmal, R.A. and Tarter, M.E.: Cumulative Polygon Address Calculation Sorting; Proceedings ACM-NCC, 20:376-384, (1965). (4.1.6)

[345] Kronrod, M.A.: An Optimal Ordering Algorithm Without a Field of Operation; Dokladi Akademia Nauk SSSR, 186:1256-1258, (1969). (4.3.2)

[346] Kruijer, H.S.M.: The Interpolated File Search Method; Informatie, 16(11):612-615, (Nov 1974). (3.2.2)

[347] Kung, H.T.: A New Upper Bound on the Complexity of Derivative Evaluation; Inf. Proc. Letters, 2(5):146-147, (Dec 1973). (6.4)

[348] Larson, P.A.: A Single-File Version of Linear Hashing with Partial Expansions; Proceedings VLDB, Mexico, 8:300-309, (Sep 1982). (3.3.14)

[349] Larson, P.A.: Analysis of Index-Sequential Files with Overflow Chaining; ACM TODS, 6(4):671-680, (Dec 1981). (3.4.3)

[350] Larson, P.A.: Analysis of Repeated Hashing; BIT, 20(1):25-32, (1980). (3.3)

[351] Larson, P.A.: Analysis of Uniform Hashing; J.ACM, 30(4):805-819, (Oct 1983). (3.3.1)

[352] Larson, P.A.: Dynamic Hashing; BIT, 18(2):184-201, (1978). (3.3.19)

[353] Larson, P.A.: Expected Worst-Case Performance of Hash Files; Computer Journal,

25(3):347-352, (Aug 1982). (3.3.2, 3.3.3, 3.3.11)

[354] Larson, P.A.: Frequency Loading and Linear Probing; BIT, 19(2):223-228, (1979). (3.3.3)

[355] Larson, P.A.: Linear Hashing with Partial Expansions; Proceedings VLDB, Montreal, 6:224-232, (1980). (3.3.14)

[356] Larson, P.A.: Performance Analysis of Linear Hashing with Partial Expansions; ACM TODS, 7(4):566-587, (Dec 1982). (3.3.14)

[357] Lee, D.T. and Wong, C.K.: Quintary Trees: A File Structure for Multi-dimensional Database System; ACM TODS, 5(3):339-353, (Sep 1980). (3.5)

[358] Lee, D.T. and Wong, C.K.: Worst-Case Analysis for Region and Partial Region Searches in Multidimensional Binary Search Trees and Balanced Quad Trees; Acta Informatica, 9(1):23-29, (1977). (3.5)

[359] Lee, K.P.: A Linear Algorithm for Copying Binary Trees Using Bounded Workspace; C.ACM, 23(3):159-162, (Mar 1980). (3.4.1)

[360] Leipala, T.: On a Generalization of Binary Search; Inf. Proc. Letters, 8(5):230-233, (June 1979). (3.2.1)

[361] Leipala, T.: On Optimal Multilevel Indexed Sequential Files; Inf. Proc. Letters, 15(5):191-195, (Dec 1982). (3.4.3)

[362] Leipala, T: On the Design of One-Level Indexed Sequential Files; Int. J of Comp and Inf Sciences, 10(3):177-186, (June 1981). (3.1.5, 3.4.3)

[363] Lesuisse, R.: Some Lessons Drawn from the History of the Binary Search Algorithm; Computer Journal, 26(2):154-163, (May 1983). (3.2.1, 2.2.2.1)

[364] Lewis, G.N., Boynton, N.J. and Burton, F.W.: Expected Complexity of Fast Search with Uniformly Distributed Data; Inf. Proc. Letters, 13(1):4-7, (Oct 1981). (3.2.2)

[365] Linial, N. and Saks, M.E.: Information Bounds Are Good for Search Problems on Ordered Data Structures; Proceedings FOCS, Tucson AZ, 24:473-475, (Nov 1983). (3)

[366] Lipton, R.J. and Dobkin, D.: Complexity Measures and Hierarchies for the Evaluation of Integers, Polynomials and N-Linear Forms; Proceedings SIGACT, Albuquerque NM, 7:1-5, (May 1975). (6.4)

[367] Lipton, R.J. and Stockmeyer, L.J.: Evaluation of Polynomials with Super-Preconditioning; Proceedings SIGACT, Hershey PA, 8:174-180, (May 1976). (6.4)

[368] Lipton, R.J., Rosenberg, A.L. and Yao, A.C-C.: External Hashing Schemes for Collection of Data Structures; J.ACM, 27(1):81-95, (Jan 1980). (3.3)

[369] Lipton, R.J.: Polynomials With 0-1 Coefficients That are Hard to Evaluate; Proceedings FOCS, Berkeley CA, 16:6-10, (Oct 1975); also SIAM J on Computing, 7(1):61-69, (Feb 1978). (6.4)

[370] Litwin, W.: Linear Hashing: A New Tool for File and Table Addressing; Proceedings VLDB, Montreal, 6:212-223, (1980). (3.3.14)

[371] Litwin, W.: Trie Hashing; Proceedings ACM-SIGMOD, 11:19-29, (Apr 1981). (3.4.4, 3.3)

[372] Litwin, W.: Virtual Hashing: A Dynamically Changing Hashing; Proceedings VLDB,

Berlin, 4:517-523, (Sep 1978). (3.3.14)

[373] Lloyd, J.W. and Ramamohanarao, K.: Partial-Match Retrieval for Dynamic Files; BIT, 22(2):150-168, (1982). (3.3.13, 3.3.14)

[374] Loeser, R.: Some Performance Tests of Quicksort and Descendants; C.ACM, 17(3):143-152, (Mar 1974). (4.1.3)

[375] Lomet, D.B.: Bounded Index Exponential Hashing; ACM TODS, 8(1):136-165, (Mar 1983). (3.3.13)

[376] Lomet, D.B.: Digital B-Trees; Proceedings VLDB, Cannes, 7:333-344, (Sep 1981). (3.4.2, 3.4.4)

[377] Lotka, A.J.: The Frequency Distribution of Scientific Production; J. of the Washington Academy of Sciences, 16(12):317-323, (1926). (I.3)

[378] Lotti, G. and Romani, F.: Application of Approximating Algorithms to Boolean Matrix Multiplication; IEEE Trans. on Computers, C-29(10):927-928, (Oct 1980). (6.3)

[379] Lowden, B.G.T.: A Note on the Oscillating Sort; Computer Journal, 20(1):92, (Feb 1977). (4.4.4)

[380] Luccio, F. and Pagli, L.: Comment on Generalized AVL Trees; C.ACM, 23(7):394-395, (July 1980). (3.4.1.3)

[381] Luccio, F. and Pagli, L.: On the Height of Height-Balanced Trees; IEEE Trans. on Computers, C-25(1):87-90, (Jan 1976). (3.4.1.3)

[382] Luccio, F. and Pagli, L.: Power Trees; C.ACM, 21(11):941-947, (Nov 1978). (3.4.1.3)

[383] Luccio, F. and Pagli, L.: Rebalancing Height Balanced Trees; IEEE Trans. on Computers, C-27(5):386-396, (May 1978). (3.4.1.3)

[384] Luccio, F.: Weighted Increment Linear Search for Scatter Tables; C.ACM, 15(12):1045-1047, (Dec 1972). (3.3.4)

[385] Lueker, G.S. and Willard, D.E.: A Data Structure for Dynamic Range Queries; Inf. Proc. Letters, 15(5):209-213, (Dec 1982). (3)

[386] Lueker, G.S.: A Data Structure for Orthogonal Range Queries; Proceedings FOCS, Ann Arbor MI, 19:28-34, (Oct 1978). (3)

[387] Lum, V.Y. and Yuen, P.S.T.: Additional Results on Key-to-Address Transform Techniques: A Fundamental Performance Study on Large Existing Formatted Files; C.ACM, 15(11):996-997, (Nov 1972). (3.3)

[388] Lum, V.Y., Yuen, P.S.T. and Dodd, M.: Key-to-address transform techniques: a fundamental performance study on large existing formatted files; C.ACM, 14(4):228-239, (1971). (3.3)

[389] Lum, V.Y.: General Performance Analysis of Key-to-Address Transformation Methods Using an Abstract File Concept; C.ACM, 16(10):603-612, (Oct 1973). (3.3)

[390] Lum, V.Y.: Multi-Attribute Retrieval with Combined Indexes; C.ACM, 13(11):660-665, (Nov 1970). (3.4.3, 3.5)

[391] Lyon, G.E.: Batch Scheduling From Short Lists; Inf. Proc. Letters, 8(2):57-59, (Feb 1979).

(3.3.8)

[392] Lyon, G.E.: Hashing with Linear Probing and Frequency Ordering; J. Res. Nat. Bureau of Standards, 83(5):445-447, (Sep 1978). (3.3.3)

[393] Lyon, G.E.: Packed Scatter Tables; C.ACM, 21(10):857-865, (Oct 1978). (3.3.9)

[394] MacCallum, I.R.: A Simple Analysis of the nth Order Polyphase Sort; Computer Journal, 16(1):16-18, (Feb 1973). (4.4.3)

[395] MacLaren, M.D.: Internal Sorting by Radix Plus Shifting; J.ACM, 13(3):404-411, (July 1966). (4)

[396] MacVeigh, D.T.: Effect of Data Representation on Cost of Sparse Matrix Operations; Acta Informatica, 7:361-394, (1977). (2.1)

[397] Maddison, J.A.T.: Fast Lookup in Hash Tables with Direct Rehashing; Computer Journal, 23(2):188-189, (Feb 1980). (3.3.8)

[398] Maier, D. and Salveter, S.C.: Hysterical B-Trees; Inf. Proc. Letters, 12(4):199-202, (Aug 1981). (3.4.2.1)

[399] Mairson, H.G.: The Program Complexity of Searching a Table; Proceedings FOCS, Tucson AZ, 24:40-47, (Nov 1983). (3.3)

[400] Makarov, O.M.: Using Duality for the Synthesis of an Optimal Algorithm Involving Matrix Multiplication; Inf. Proc. Letters, 13(2):48-49, (Nov 1981). (6.3)

[401] Malcolm, W.D.: String Distribution for the Polyphase Sort; C.ACM, 6(5):217-220, (May 1963). (4.4.3)

[402] Mallach, E.G.: Scatter Storage Techniques: A Unifying Viewpoint and a Method for Reducing Retrieval Times; Computer Journal, 20(2):137-140, (May 1977). (3.3.8)

[403] Maly, K.: A Note on Virtual Memory Indexes; C.ACM, 21(9):786-787, (Sep 1978). (3.4.2)

[404] Maly, K.: Compressed Tries; C.ACM, 19(7):409-415, (July 1976). (3.4.4)

[405] Manacher, G.K.: Significant Improvements to the Hwang-Lin Merging Algorithm; J.ACM, 26(3):434-440, (July 1979). (4.3.3)

[406] Manacher, G.K.: The Ford-Johnson Sorting Algorithm is Not Optimal; J.ACM, 26(3):441-456, (July 1979). (4.1)

[407] Manker, H.H.: Multiphase Sorting; C.ACM, 6(5):214-217, (May 1963). (4.4.3)

[408] Martin, W.A. and Ness, D.N.: Optimizing Binary Trees Grown with a Sorting Algorithm; C.ACM, 15(2):88-93, (Feb 1972). (3.4.1.6)

[409] Martin, W.A.: Sorting; Computing Surveys, 3(4):147-174, (Dec 1971). (4.1, 4.4)

[410] Maruyama, K. and Smith, S.E.: Analysis of Design Alternatives for Virtual Memory Indexes; C.ACM, 20(4):245-254, (Apr 1977). (3.4.3)

[411] Maurer, H.A., Ottmann, Th. and Six, H.W.: Implementing Dictionaries Using Binary Trees of Very Small Height; Inf. Proc. Letters, 5(1):11-14, (May 1976). (3.4.2.3)

[412] Maurer, W.D. and Lewis, T.E.: Hash table methods; ACM C. Surveys, 7(1):5-19, (Mar 1975). (3.3)

[413] Maurer, W.D.: An Improved Hash Code for Scatter Storage; C.ACM, 11(1):35-38, (Jan 1968). (3.3, 3.3.5)

[414] McAllester, R.L.: Polyphase Sorting with Overlapped Rewind; C.ACM, 7(3):158-159, (Mar 1964). (4.4.3)

[415] McCabe, J.: On serial files with relocatable records; Operations Research, 13(4):609-618, (1965). (3.1.2)

[416] McCreight, E.: Pagination of B*-trees with variable-length records; C.ACM, 20(9):670-674, (Sep 1977). (3.4.2)

[417] McCulloch, C.M.: Quickshunt - A Distributive Sorting Algorithm; Computer Journal, 25(1):102-104, (Feb 1982). (4.2.4, 4.4)

[418] McKellar, A.C. and Wong, C.K.: Bounds on Algorithms for String Generation; Acta Informatica, 1(4):311-319, (1972). (4.4.0.2)

[419] McKellar, A.C. and Wong, C.K.: Dynamic Placement of Records in Linear Storage; J.ACM, 25(3):421-434, (July 1978). (3.1)

[420] Mehlhorn, K. and Overmars, M.H.: Optimal Dynamization of Decomposable Searching Problems; Inf. Proc. Letters, 12(2):93-98, (Apr 1981). (2.2)

[421] Mehlhorn, K.: A Best Possible Bound for the Weighted Path Length of Binary Search Trees; SIAM J on Computing, 6(2):235-239, (June 1977). (3.4.1.6)

[422] Mehlhorn, K.: A Partial Analysis of Height-Balanced Trees Under Random Insertions and Deletions; SIAM J on Computing, 11(4):748-760, (Nov 1982). (3.4.1.3, 3.4.2.1, 3.4.2.3)

[423] Mehlhorn, K.: Dynamic Binary Search; SIAM J on Computing, 8(2):175-198, (May 1979). (3.4.1.6, 3.4.4)

[424] Mehlhorn, K.: Nearly Optimal Binary Search Trees; Acta Informatica, 5:287-295, (1975). (3.4.1.6)

[425] Mehlhorn, K.: On the Program Size of Perfect and Universal Hash Functions; Proceedings FOCS, Chicago IL, 23:170-175, (Oct 1982). (3.3)

[426] Mehlhorn, K.: Sorting Presorted Files; Proceedings 4th GI Conference, Lecture Notes in Comp. Science 67:199-212, (1979). (4.1)

[427] Meijer, H. and Akl, S.G.: The Design and Analysis of a New Hybrid Sorting Algorithm; Inf. Proc. Letters, 10(4):213-218, (July 1980). (4.1.1, 4.1.8)

[428] Meir, A. and Moon, J.W.: On the Altitude of Nodes in Random Trees; Canad J Math, 30(5):997-1015, (1978). (3.4)

[429] Melville, R. and Gries, D.: Controlled Density Sorting; Inf. Proc. Letters, 10(4):169-172, (July 1980). (4.1.2, 4.1.7)

[430] Mendelson, H. and Yechiali, U.: A New Approach to the Analysis of Linear Probing Schemes; J.ACM, 27(2):474-483, (July 1980). (3.3.3)

[431] Mendelson, H. and Yechiali, U.: Performance Measures for Ordered Lists in Random-Access Files; J.ACM, 26(4):654-667, (Oct 1979). (3.3)

[432] Mescheder, B.: On the Number of Active *-Operations Needed to Compute the Discrete

Fourier Transform; Acta Informatica, 13(4):383-408, (1980). (6)

[433] Miller, R., Pippenger, N., Rosenberg, A. and Snyder, L.: Optimal 2-3 trees; SIAM J on Computing, 8(1):42-59, (Feb 1979). (3.4.2.1)

[434] Miyakawa, M., Yuba, T., Sugito, Y. and Hoshi, M.: Optimum Sequence Trees; SIAM J on Computing, 6(2):201-234, (June 1977). (3.4.4)

[435] Moenk, R. and Borodin, A.: Fast Modular Transforms Via Division; Proceedings FOCS, College Park Md, 13:90-96, (Oct 1972). (6.4)

[436] Monard, M.C.: Design and Analysis of External Quicksort Algorithms; PhD Dissertation, PUC University, Rio de Janeiro Brazil, (Feb 1980). (4.4.5)

[437] Morris, R.: Counting Large Numbers of Events in Small Registers; C.ACM, 21(10):840-842, (Oct 1978). (6.1)

[438] Morris, R.: Scatter Storage Techniques; C.ACM, 11(1):38-44, (Jan 1968). (3.3.2, 3.3.3, 3.3.10, 3.3.11)

[439] Motoki, T.: A Note on Upper Bounds for the Selection Problem; Inf. Proc. Letters, 15(5):214-219, (Dec 1982). (5.2)

[440] Motzkin, D.: A Stable Quicksort; Software - Practice and Experience, 11:607-611, (1981). (4.2.2)

[441] Motzkin, D.: Meansort; C.ACM, 26(4):250-251, (Apr 1983). (4.1.3)

[442] Motzkin, T.S.: Evaluation of Polynomials and Evaluation of Rational Functions; Bull of Amer Math Soc, 61:163, (1965). (6.4)

[443] Mullin, J.K.: An Improved Index Sequential Access Method Using Hashed Overflow; C.ACM, 15(5):301-307, (May 1972). (3.4.3)

[444] Mullin, J.K.: Tightly Controlled Linear Hashing Without Separate Overflow Storage; BIT, 21(4):390-400, (1981). (3.3.14)

[445] Munro, J.I. and Paterson, M.S.: Selection and Sorting with Limited Storage; Proceedings FOCS, Ann Arbor MI, 19:253-258, (Oct 1978); also Theoretical Computer Science, 12:315-323, (1980). (4.4, 5.2)

[446] Munro, J.I. and Poblete, P.V.: A Discipline for Robustness or Storage Reduction in Binary Search Trees; Proceedings ACM PODS, Atlanta GA, 2:70-75, (Mar 1983). (3.4.1)

[447] Munro, J.I. and Spira, P.M.: Sorting and Searching in Multisets; SIAM J on Computing, 5(1):1-8, (Mar 1976). (4.2)

[448] Munro, J.I. and Suwanda, H.: Implicit Data Structures for Fast Search and Update; Proceedings SIGACT, Atlanta GA, 11:108-117, (Apr 1979); also JCSS, 21(2):236-250, (Oct 1980).(3)

[449] Murphy, L.J.: Lotka's Law in the Humanities; J. of the American Society of Information Science, 24(6):461-462, (1973). (I.3)

[450] Nakamura, T. and Mizoguchi, T.: An Analysis of Storage Utilization Factor in Block Split Data Structuring Scheme; Proceedings VLDB, Berlin, 4:489-495, (Sep 1978). (3.4.3)

[451] Nat, M. van der: A Fast Sorting Algorithm, a Hybrid of Distributive and Merge Sorting;

Inf. Proc. Letters, 10(3):163-167, (Apr 1980). (4.2.5)

[452] Nat, M. van der: Binary Merging by Partitioning; Inf. Proc. Letters, 8(2):72-75, (Feb 1979). (4.3)

[453] Nat, M. van der: On Interpolation Search; C.ACM, 22(12):681, (Dec 1979). (3.2.2)

[454] Nevalainen, O. and Teuhola, J.: Priority Queue Administration by Sublist Index; Computer Journal, 22(3):220-225, (Mar 1979). (5.1.1)

[455] Nevalainen, O. and Teuhola, J.: The Efficiency of Two Indexed Priority Queue Algorithms; BIT, 18(3):320-333, (1978). (5.1.2)

[456] Nicklas, B.M. and Schlageter, G.: Index Structuring in Inverted Data Bases by Tries; Computer Journal, 20(4):321-324, (Nov 1977). (3.4.4)

[457] Nievergelt, J. and Reingold, E.M.: Binary Search Trees of Bounded Balance; SIAM J on Computing, 2(1):33-43, (1973). (3.4.1.4)

[458] Nievergelt, J.: Binary Search Trees and File Organization; ACM C. Surveys, 6(3):195-207, (Sep 1974). (3.4.1)

[459] Noshita, K.: Median Selection of 9 Elements in 14 Comparisons; Inf. Proc. Letters, 3(1):8-12, (July 1974). (5.2)

[460] Nozaki, A.: A Note on the Complexity of Approximative Evaluation of Polynomials; Inf. Proc. Letters, 9(2):73-75, (Aug 1979). (6.4)

[461] Nozaki, A.: Sorting Using Networks of Deques; JCSS, 19(3):309-315, (Dec 1979). (4.2)

[462] Nozaki, A.: Two Entropies of a Generalized Sorting Problem; JCSS, 7(5):615-621, (Oct 1973). (4, 5.2)

[463] O'Dunlaing, C. and Yap, C.K.: Generic Transformation of Data Structures; Proceedings FOCS, Chicago IL, 23:186-195, (Oct 1982). (2.1)

[464] Odlyzko, A.M.: Periodic Oscillations of Coefficients of Power Series that Satisfy Functional Equations; Advances in Mathematics, to appear. (3.4.2)

[465] Olivie, H.J.: On a Relationship Between 2-3 Brother Trees and Dense Ternary Trees; Int. J Computer Math, A8:233-245, (1980). (3.4.2.1)

[466] Olivie, H.J.: On Random Son-trees; Int. J Computer Math, 9:287-303, (1981). (3.4.2.3)

[467] Olivie, H.J.: On the Relationship Between Son-trees and Symmetric Binary B-trees; Inf. Proc. Letters, 10(1):4-8, (Feb 1980). (3.4.2.2, 3.4.2.3)

[468] Orenstein, J.A.: Multidimensional Tries Used for Associative Searching; Inf. Proc. Letters, 14(4):150-157, (June 1982). (3.4.4, 3.5)

[469] Ottmann, Th. and Stucky, W.: Higher Order Analysis of Random 1-2 Brother Trees; BIT, 20(3):302-314, (1980). (3.4.2.3)

[470] Ottmann, Th. and Wood, D.: 1-2 Brother Trees or AVL Trees Revisited; Computer Journal, 23(3):248-255, (Aug 1980). (3.4.1.3, 3.4.2.3)

[471] Ottmann, Th. and Wood, D.: A Comparison of Iterative and Defined Classes of Search Trees; Int. J of Comp and Inf Sciences, 11(3):155-178, (June 1982). (3.4.1, 3.4.2)

[472] Ottmann, Th. and Wood, D.: Deletion in One-Sided Height-Balanced Search Trees; Int. J

Computer Math, 6(4):265-271, (1978). (3.4.1.3)

[473] Ottmann, Th., Six, H.W. and Wood, D.: On the Correspondence Between AVL Trees and Brother Trees; Computing, 23(1):43-54, (1979). (3.4.2.3, 3.4.1.3)

[474] Ottmann, Th., Six, H.W. and Wood, D.: One-Sided k-Height-Balanced Trees; Computing, 22(4):283-290, (1979). (3.4.1.3)

[475] Ottmann, Th., Six, H.W. and Wood, D.: Right Brother Trees; C.ACM, 21(9):769-776, (Sep 1978). (3.4.2.3)

[476] Ottmann, Th., Six, H.W. and Wood, D.: The Implementation of Insertion and Deletion Algorithms for 1-2 Brother Trees; Computing, 26:369-378, (1981). (3.4.2.3)

[477] Ouksel, M. and Scheuermann, P.: Multidimensional B-Trees: Analysis of Dynamic Behavior; BIT, 21(4):401-418, (1981). (3.4.2, 3.5)

[478] Ouksel, M. and Scheuermann, P.: Storage Mappings for Multidimensional Linear Dynamic Hashing; Proceedings ACM PODS, Atlanta GA, 2:90-105, (Mar 1983). (3.3.19)

[479] Overholt, K.J.: Efficiency of the Fibonacci Search Method; BIT, 13(1):92-96, (1973). (3.2)

[480] Overholt, K.J.: Optimal Binary Search Methods; BIT, 13(1):84-91, (1973). (3.2.1)

[481] Overmars, M.H. and vanLeeuwen, J.: Dynamic Multidimensional Data Structures Based on Quad- and K-D Trees; Acta Informatica, 17(3):267-285, (1982). (2.2, 3.5.1, 3.5.2)

[482] Overmars, M.H. and vanLeeuwen, J.: Dynamizations of Decomposable Searching Problems Yielding Good Worst-Case Bounds; Lecture Notes in Computer Science, Springer-Verlag 104:224-233, (1981). (2.2)

[483] Overmars, M.H. and vanLeeuwen, J.: Some Principles for Dynamizing Decomposable Searching Problems; Inf. Proc. Letters, 12(1):49-53, (Feb 1981). (2.2)

[484] Overmars, M.H. and vanLeeuwen, J.: Two General Methods for Dynamizing Decomposable Searching Problems; Computing, 26(2):155-166, (1981). (2.2)

[485] Overmars, M.H. and vanLeeuwen, J.: Worst-Case Optimal Insertion and Deletion Methods for Decomposable Searching Problems; Inf. Proc. Letters, 12(4):168-173, (Aug 1981). (2.2)

[486] Overmars, M.H.: Dynamization of Order Decomposable Set Problems; J. of Algorithms, 2(3):245-260, (Sep 1981). (2.2)

[487] Palmer, E.M., Rahimi, M.A. and Robinson, R.W.: Efficiency of a Binary Comparison Storage Technique; J.ACM, 21(3):376-384, (July 1974). (3.4.1.1)

[488] Pan, V.Y.: A Unified Approach to the Analysis of Bilinear Algorithms; J. of Algorithms, 2(3):301-310, (Sep 1981). (6.3)

[489] Pan, V.Y.: Computational Complexity of Computing Polynomials Over the Field of Real and Complex Numbers; Proceedings SIGACT, San Diego CA, 10:162-172, (May 1978). (6.4)

[490] Pan, V.Y.: New Combinations of Methods for the Acceleration of Matrix Multiplication; Comput Math with Applic, 7:73-125, (1981). (6.3)

[491] Pan, V.Y.: New Fast Algorithms for Matrix Operations; SIAM J on Computing, 9(2):321-

342, (May 1980). (6.3)

[492] Pan, V.Y.: New Methods for the Acceleration of Matrix Multiplication; Proceedings FOCS, San Juan PR, 20:28-38, (Oct 1979). (6.3)

[493] Pan, V.Y.: Strassen's Algorithm is not Optimal Trilinear Technique of Aggregating, Uniting and Canceling for Constructing Fast Algorithms for Matrix Operations; Proceedings FOCS, Ann Arbor MI, 19:166-176, (Oct 1978). (6.3)

[494] Pan, V.Y.: The Additive and Logical Complexities of Linear and Bilinear Arithmetic Algorithms; J. of Algorithms, 4(1):1-34, (Mar 1983). (6.3)

[495] Pan, V.Y.: The Bit-Complexity of Arithmetic Algorithms; J. of Algorithms, 2(2):144-163, (June 1981). (6.4)

[496] Papadimitriou, C.H. and Bernstein, P.A.: On the Performance of Balanced Hashing Functions When Keys are Not Equiprobable; ACM TOPLAS, 2(1):77-89, (Jan 1980). (3.3)

[497] Peltola, E. and Erkio, H.: Insertion Merge Sorting; Inf. Proc. Letters, 7(2):92-99, (Feb 1978). (4.2.1, 4.2.5)

[498] Perl, Y. and Reingold, E.M.: Understanding the Complexity of Interpolation Search; Inf. Proc. Letters, 6(6):219-221, (Dec 1977). (3.2.2)

[499] Perl, Y., Itai, A. and Avni, H.: Interpolation Search - A Log Log N Search; C.ACM, 21(7):550-553, (July 1978). (3.2.2)

[500] Peters, J.G. and Kritzinger, P.S.: Implementation of Samplesort: A Minimal Storage Tree Sort; BIT, 15(1):85-93, (1975). (4.1.3)

[501] Peterson, W.W.: Addressing for Random-Access Storage; IBM J Res. Development, 1(4):130-146, (Apr 1957). (3.2, 3.3)

[502] Pippenger, N.: Computational Complexity in Algebraic Functions Fields; Proceedings FOCS, San Juan PR, 20:61-65, (Oct 1979). (6.2)

[503] Pippenger, N.: On the Application of Coding Theory to Hashing; IBM J Res. Development, 23(2):225-226, (Mar 1979). (3.3)

[504] Pippenger, N.: On the Evaluation of Powers and Monomials; SIAM J on Computing, 9(2):230-250, (May 1980). (6.2)

[505] Pippenger, N.: On the Evaluation of Powers and Related Problems; Proceedings FOCS, Houston TX, 17:258-263, (Oct 1976). (6.2)

[506] Pohl, I.: Minimean Optimality in Sorting Algorithms; Proceedings FOCS, Berkeley CA, 16:71-74, (Oct 1975). (4.1, 5.1)

[507] Pooch, U.W. and Nieder, A.: A Survey of Indexing Techniques for Sparse Matrices; ACM C. Surveys, 5(2):109-133, (June 1973). (2.1)

[508] Pope, A.: Bradford's Law and the Periodical Literature of Information Sciences; J. of the Association of Information Sciences, 26(4):207-213, (1975). (I.2)

[509] Porter, T. and Simon, I.: Random Insertion into a Priority Queue Structure; IEEE Trans. Software Engineering, 1(3):292-298, (Sep 1975). (5.1.3)

[510] Power, L.R.: Internal Sorting Using a Mimimal Tree Merge Strategy; ACM TOMS,

6(1):68-79, (Mar 1980). (4.2)

[511] Pratt, V. and Yao, F.: On Lower Bounds for Computing the ith Largest Element; Proceedings SWAT (FOCS), Iowa City IO, 14:70-81, (Oct 1973). (5.2)

[512] Pratt, V.: The Power of Negative Thinking in Multiplying Boolean Matrices; Proceedings SIGACT, Seattle WA, 6:80-83, (Apr 1974). (6.3)

[513] Price, C.E.: Table Lookup Techniques; ACM C. Surveys, 3(2):56-58, (June 1971). (3.2)

[514] Probert, R.L.: An Extension of Computational Duality to Sequences of Bilinear Computations; SIAM J on Computing, 7(1):91-98, (Feb 1978). (6.3)

[515] Probert, R.L.: Commutativity, Non-Commutativity and Bilinearity; Inf. Proc. Letters, 5(2):46-49, (June 1976). (6.3)

[516] Probert, R.L.: On the Additive Complexity of Matrix Multiplication; SIAM J on Computing, 5(2):187-203, (June 1976). (6.3)

[517] Proskurowski, A.: On the Generation of Binary Trees; J.ACM, 27(1):1-2, (Jan 1980). (3.4.1)

[518] Quittner, P., Csoka, S., Halasz, S., Kotsis, D. and Varnai, K.: Comparison of Synomym Handling and Bucket Organization Methods; C.ACM, 24(9):579-583, (Sep 1981). (3.3.3, 3.3.11)

[519] Radhakrishnan, T. and Kernizan, R.: Lotka's Law and Computer Science Literature; J Am. Society for Information Science, 30(1):51-54 (Jan 1979). (I.3)

[520] Radke, C.E.: The Use of Quadratic Residue Research; C.ACM, 13(2):103-105, (Feb 1970). (3.3.5)

[521] Raghavan, V.V. and Yu, C.T.: A Note on a Multidimensional Searching Problem; Inf. Proc. Letters, 6(4):133-135, (Aug 1977). (3.5)

[522] Raiha, K.J. and Zweben, S.H.: An Optimal Insertion Algorithm for One-Sided Height-Balanced Binary Search Trees; C.ACM, 22(9):508-512, (Sep 1979). (3.4.1.3)

[523] Ramamohanarao, K. and Lloyd, J.W.: Dynamic Hashing Schemes; Computer Journal, 25(4):478-485, (Nov 1982). (3.3.14)

[524] Regnier, M.: On the Average Height of Trees in Digital Search and Dynamic Hashing; Inf. Proc. Letters, 13(2):64-66, (Nov 1981). (3.4.4, 3.3.19)

[525] Reingold, E.M.: A note on 3-2 trees; Fibonacci Quarterly, 17(2):151-157 (Apr 1979). (3.4.2.1)

[526] Reiser, A.: A Linear Selection Algorithm for Sets of Elements with Weights; Inf. Proc. Letters, 7(3):159-162, (Apr 1978). (5.2)

[527] Remy, J.L.: Construction Evaluation et Amelioration Systematiques de Structures de Donnees; RAIRO Informatique Theorique, 14(1):83-118, (1980). (2.2)

[528] Revah, L.: On the Number of Multiplications/Divisions Evaluating a Polynomial with Auxiliary Functions; SIAM J on Computing, 4(3):381-392, (Sep 1975). (6.4)

[529] Rivest, R.L. and Wiele, J.P. van de: An $\Omega((n/lg\,n)^{1/2})$ Lower Bound on the Number of Additions Necessary to Compute 0-1 Polynomials Over the Ring of Integer Polynomials;

Inf. Proc. Letters, 8(4):178-180, (Apr 1979). (6.4)

[530] Rivest, R.L.: On Hash-Coding Algorithms for Partial-Match Retrieval; Proceedings FOCS, New Orleans LA, 15:95-103, (Oct 1974). (3.3)

[531] Rivest, R.L.: On Self-Organizing Sequential Search Heuristics; Proceedings FOCS, New Orleans LA, 15:122-126, (Oct 1974); also C.ACM, 19(2):63-67, (Feb 1976). (3.1.2, 3.1.3)

[532] Rivest, R.L.: Optimal Arrangement of Keys in a Hash Table; J.ACM, 25(2):200-209, (Apr 1978). (3.3.8)

[533] Rivest, R.L.: Partial-Match Retrieval Algorithms; SIAM J on Computing, 5(1):19-50, (Mar 1976). (3)

[534] Robinson, J.T.: The k-d-B-tree: A Search Structure for Large Multidimensional Dynamic Indexes; Proceedings ACM-SIGMOD, :10-18, (Apr 1981). (3.4.2, 3.5.2)

[535] Robson, J.M.: The Height of Binary Search Trees; Australian Computer J, 11(4):151-153, (Nov 1979). (3.4.1.1)

[536] Rohrich, J.: A Hybrid of Quicksort with O(n log n) Worst Case Complexity; Inf. Proc. Letters, 14(3):119-123, (May 1982). (4.1.3)

[537] Romani, F.: Some Properties of Disjoint Sums of Tensors Related to Matrix Multiplication; SIAM J on Computing, 11(2):263-267, (May 1982). (6.3)

[538] Rosenberg, A.L. and Snyder, L.: Minimal comparison 2-3 trees; SIAM J on Computing, 7(4):465-480, (Nov 1978). (3.4.2.1)

[539] Rosenberg, A.L. and Snyder, L.: Time- and Space-Optimality in B-Trees; ACM TODS, 6(1):174-193, (Mar 1981). (3.4.2)

[540] Rosenberg, A.L. and Stockmeyer, L.J.: Hashing Schemes for Extendible Arrays; Proceedings SIGACT, Albuquerque NM, (May 1975), 7:159-166; also J.ACM, 24(2):199-221, (Apr 1977). (3.3)

[541] Rosenberg, A.L. and Stockmeyer, L.J.: Storage Schemes for Boundedly Extendible Arrays; Acta Informatica, 7:289-303, (1977). (2.1)

[542] Rosenberg, A.L., Stockmeyer, L.J. and Snyder, L.: Uniform Data Encodings; Theoretical Computer Science, 11(2):145-165, (1980). (2.1)

[543] Rosenberg, A.L., Wood, D. and Galil, Z.: Storage Representations for Tree-Like Data Structures; Proceedings SIGACT, Atlanta GA, 11:99-107, (Apr 1979); also Math Systems Theory, 13(2):105-130, (1979). (2.1)

[544] Rosenberg, A.L.: Allocating Storage for Extendible Arrays; J.ACM, 21(4):652-670, (Oct 1974). (2.1)

[545] Rosenberg, A.L.: Data Encodings and their Costs; Acta Informatica, 9(3):273-292, (1978). (2.1)

[546] Rosenberg, A.L.: Encoding Data Structures in Trees; J.ACM, 26(4):668-689, (Oct 1979). (3.4)

[547] Rosenberg, A.L.: Managing Storage for Extendible Arrays; Proceedings SIGACT, Seattle WA, 6:297-302, (Apr 1974); also SIAM J on Computing, 4(3):287-306, (Sep 1975). (2.1)

[548] Rosenberg, A.L.: On Uniformly Inserting One Data Structure into Another; C.ACM, 24(2):88-90, (Feb 1981). (2.1, 2.2)

[549] Rotem, D. and Varol, Y.L.: Generation of Binary Trees from Ballot Sequences; J.ACM, 25(3):396-404, (July 1978). (3.4.1)

[550] Rothnie, J.B. and Lozano, T.: Attribute Based File Organization in a Paged Memory Environment; C.ACM, 17(2):63-69, (Feb 1974). (3.3, 3.4)

[551] Ruskey, F. and Hu, T.C.: Generating Binary Trees Lexicographically; SIAM J on Computing, 6(4):745-758, (Dec 1977). (3.4.1)

[552] Ruskey, F.: Generating t-Ary Trees Lexicographically; SIAM J on Computing, 7(4):424-439, (Nov 1978). (3.4.6)

[553] Samadi, B.: B-trees in a system with multiple views; Inf. Proc. Letters, 5(4):107-112, (Oct 1976). (3.4.2)

[554] Samet, H.: A Quadtree Medial Axis Transform; C.ACM, 26(9):680-693, (Sep 1983). (3.5.1)

[555] Samet, H.: Deletion in Two-Dimensional Quad Trees; C.ACM, 23(12):703-710, (Dec 1980). (3.5.1)

[556] Samson, W.B. and Davis, R.H.: Search Times Using Hash Tables for Records with Non-Unique Keys; Computer Journal, 21(3):210-214, (Aug 1978). (3.3.5)

[557] Samson, W.B.: Hash Table Collision Handling on Storage Devices with Latency; Computer Journal, 24(2):130-131, (May 1981). (3.3.3, 3.3.4)

[558] Santoro, N.: Extending the Four Russians' Bound to General Matrix Multiplication; Inf. Proc. Letters, 10(2):87-88, (Mar 1980). (6.3)

[559] Santoro, N.: Full Table Search by Polynomial Functions; Inf. Proc. Letters, 5(3):72-74, (Aug 1976). (3.3.5)

[560] Sarwate, D.V.: A Note on Universal Classes of Hash Functions; Inf. Proc. Letters, 10(1):41-45, (Feb 1980). (3.3)

[561] Savage, J.E.: An Algorithm for the Computation of Linear Forms; SIAM J on Computing, 3(2):150-158, (June 1974). (6.3, 6.4)

[562] Saxe, J.B. and Bentley, J.L.: Transforming Static Data Structures to Dynamic Data Structures; Proceedings FOCS, San Juan PR, 20:148-168 (Oct 1979). (2.2)

[563] Saxe, J.B.: On the Number of Range Queries in k-Space; Discr App Math, 1(3):217-225, (1979). (3.5)

[564] Schachtel, G.: A Noncommutative Algorithm for Multiplying 5 x 5 Matrices Using 103 Multiplications; Inf. Proc. Letters, 7(4):180-182, (June 1978). (6.3)

[565] Schay, G. and Spruth, W.G.: Analysis of a File Addressing Method; C.ACM, 5(8):459-462, (Aug 1962). (3.3.3)

[566] Schlumberger, M. and Vuillemin, J.: Optimal Disk Merge Patterns; Acta Informatica, 3(1):25-35, (1973). (4.3, 4.4)

[567] Schnorr, C.P. and Wiele, J.P. van de: On the Additive Complexity of Polynomials;

Theoretical Computer Science, 10(1):1-18, (1980). (6.4)

[568] Schnorr, C.P.: How Many Polynomials Can be Approximated Faster than they can be Evaluated?; Inf. Proc. Letters, 12(2):76-78, (Apr 1981). (6.4)

[569] Scholl, M.: New File Organizations Based on Dynamic Hashing; ACM TODS, 6(1):194-211, (Mar 1981). (3.3.19)

[570] Schonhage, A., Paterson, M. and Pippenger, N.: Finding the Median; JCSS, 13(2):184-199, (Oct 1976). (5.2)

[571] Schonhage, A.: Fast Multiplication of Polynomials Over Fields of Characteristic 2; Acta Informatica, 7:395-398, (1977). (6)

[572] Schonhage, A.: Partial and Total Matrix Multiplication; SIAM J on Computing, 10(3):434-455, (Aug 1981). (6.3)

[573] Schoor, A.: Fast Algorithm for Sparse Matrix Multiplication; Inf. Proc. Letters, 15(2):87-89, (Sep 1982). (6.3)

[574] Schulte Monting, J.: Merging of 4 or 5 Elements with n Elements; Theoretical Computer Science, 14(1):19-37, (1981). (4.3.3)

[575] Scowen, R.S.: Algorithm 271, Quickersort; C.ACM, 8(11):669-670, (Nov 1965); also 9(5):354, (May 1966). (4.1.3)

[576] Sedgewick, R.: A New Upper Bound for Shellsort; J. of Algorithms, to appear. (4.1.4)

[577] Sedgewick, R.: Data Movement in Odd-Even Merging; SIAM J on Computing, 7(3):239-272, (Aug 1978). (4.2, 4.3)

[578] Sedgewick, R.: Implementing Quicksort Programs; C.ACM, 21(10):847-856, (Oct 1978). (4.1.3)

[579] Sedgewick, R.: Quicksort With Equal Keys; SIAM J on Computing, 6(2):240-267, (June 1977). (4.1.3)

[580] Sedgewick, R.: The Analysis of Quicksort Programs; Acta Informatica, 7:327-355, (1977). (4.1.3)

[581] Severance, D.G. and Duhne, R.: A Practitioner's Guide to Addressing Algorithms; C.ACM, 19(6):314-326, (June 1976). (3.3)

[582] Severance, D.G.: Identifier Search Mechanisms: A Survey and Generalized Model; ACM C. Surveys, 6(3):174-194, (Sep 1974). (3)

[583] Shaw, M. and Traub, J.F.: On the Number of Multiplications for the Evaluation of a Polynomial and Some of its Derivatives; Proceedings FOCS, College Park Md, 13:105-107, (Oct 1972); also J.ACM, 21(1):161-167, (Jan 1974). (6.4)

[584] Shaw, M. and Traub, J.F.: Selection of Good Algorithms from a Family of Algorithms for Polynomial Derivative Evaluation; Inf. Proc. Letters, 6(5):141-145, (Oct 1977). (6.4)

[585] Sheil, B.A.: Median Split Trees: A Fast Lookup Technique for Frequently Ocurring Keys; C.ACM, 21(11):947-958, (Nov 1978). (3.4.1.6)

[586] Shell, D.L.: A High-Speed Sorting Procedure; C.ACM, 2(7):30-32, (July 1959). (4.1.4)

[587] Shell, D.L.: Optimizing the Polyphase Sort; C.ACM, 14(11):713-719, (Nov 1971). (4.4.3)

[588] Shneiderman, B.: A Model for Optimizing Indexed File Structures; Int. J of Comp and Inf Sciences, 3(1):93-103, (Mar 1974). (3.4.3)

[589] Shneiderman, B.: Jump Searching: A Fast Sequential Search Technique; C.ACM, 21(10):831-834, (Oct 1978). (3.1.5)

[590] Silva Filho, Y.V.: Average Case Analysis of Region Search in Balanced k-d Trees; Inf. Proc. Letters, 8(5):219-223, (June 1979). (3.5.2)

[591] Silva Filho, Y.V.: Optimal Choice of Discriminators in a Balanced k-d Binary Search Tree; Inf. Proc. Letters, 13(2):67-70, (Nov 1981). (3.5.2)

[592] Singleton, R.C.: An Efficient Algorithm for Sorting with Minimal Storage; C.ACM, 12(3):185-187, (Mar 1969). (4.1.3)

[593] Six, H.W.: Improvement of the m-way Search Procedure; Angewandte Informatik, 15(1):79-83, (Feb 1973). (3.1.5)

[594] Sleator, D.D. and Tarjan, R.E.: Self-Adjusting Binary Trees; Proceedings SIGACT, Boston, 15:235-245, (Apr 1983). (3.4.1.6, 5.1.6)

[595] Snyder, L.: On B-Trees Re-Examined; C.ACM, 21(7):594, (July 1978). (3.4.2)

[596] Snyder, L.: On Uniquely Represented Data Structures; Proceedings FOCS, Providence RI, 18:142-146, (Oct 1977). (3.4)

[597] Sobel, S.: Oscillating Sort - a New Sort Merging Technique; J.ACM, 9:372-374, (1962). (4.4.4)

[598] Solomon, M. and Finkel, R.A.: A Note on Enumerating Binary Trees; J.ACM, 27(1):3-5, (Jan 1980). (3.4.1)

[599] Sprugnoli, R.: On the Allocation of Binary Trees to Secondary Storage; BIT, 21(3):305-316, (1981). (3.4.1.1)

[600] Sprugnoli, R.: Perfect Hashing Functions: A Single Probe Retrieving Method for Static Sets; C.ACM, 20(11):841-850, (Nov 1977). (3.3)

[601] Stephenson, C.J.: A Method for Constructing Binary Search Trees by Making Insertions at the Root; Int. J of Comp and Inf Sciences, 9(1):15-29, (Feb 1980). (3.4.1)

[602] Stockemyer, P.K. and Yao, F.F.: On the Optimality of Linear Merge; SIAM J on Computing, 9(1):85-90, (Feb 1980). (4.3.3)

[603] Stockmeyer, L.J.: The Complexity of Approximate Counting; Proceedings SIGACT, Boston Mass, 15:118-126, (Apr 1983). (6.1)

[604] Strassen, V.: Gaussian Elimination is not Optimal; Numer Math, 13:354-356, (1969). (6.3)

[605] Strassen, V.: Polynomials with Rational Coefficients Which are Hard to Compute; SIAM J on Computing, 3(2):128-149, (June 1974). (6.4)

[606] Strong, H.R., Markowsky, G. and Chandra, A.K.: Search Within a Page; J.ACM, 26(3):457-482, (July 1979). (3.4.1, 3.4.2, 3.4.3)

[607] Szwarcfiter, J.L. and Wilson, L.B.: Some Properties of Ternary Trees; Computer Journal, 21(1):66-72, (Feb 1978). (3.4.6, 4.2.8)

[608] Tai, K.C. and Tharp, A.L.: Computed Chaining A Hybrid of Direct and Open Addressing;

Proceedings ACM-NCC, 49:275-282, (1980). (3.3, 3.3.10)

[609] Tamminen, M.: Analysis of N-Trees; Inf. Proc. Letters, 16(3):131-137, (Apr 1983). (3.4.2)

[610] Tamminen, M.: Extendible Hashing with Overflow; Inf. Proc. Letters, 15(5):227-232, (Dec 1982). (3.3.13)

[611] Tamminen, M.: Order Preserving Extendible Hashing and Bucket Tries; BIT, 21(4):419-435, (1981). (3.3.13, 3.4.4)

[612] Tan, K.C.: On Foster's Information Storage and Retrieval Using AVL Trees; C.ACM, 15(9):843, (Sep 1972). (3.4.1.3)

[613] Tan, K.P. and Hsu, L.S.: Block Sorting of a Large File in External Storage by a 2-Component Key; Computer Journal, 25(3):327-330, (Aug 1982). (4.4)

[614] Tanner, R.M.: Minimean Merging and Sorting: An Algorithm; SIAM J on Computing, 7(1):18-38, (Feb 1978). (4.3, 4.2)

[615] Tarjan, R.E. and Yao, A.C-C.: Storing a Sparse Table; C.ACM, 22(11):606-611, (Nov 1979). (2.1)

[616] Tarjan, R.E.: Sorting Using Networks of Queues and Stacks; J.ACM, 18(2):341-346, (Apr 1972). (4.2)

[617] Tarjan, R.E.: Updating a Balanced Search Tree in O(1) Rotations; Inf. Proc. Letters, 16(5):253-257, (June 1983). (3.4.2.2)

[618] Tarter, M.E. and Kronmal, R.A.: Non-Uniform Key Distribution and Address Calculation Sorting; Proceedings ACM-NCC, 21:331-337, (1966). (4.1.6, 4.2.3)

[619] Tenenbaum, A.M. and Nemes, R.M.: Two Spectra of Self-Organizing Sequential Algorithms; SIAM J on Computing, 11(3):557-566, (Aug 1982). (3.1.2)

[620] Tenenbaum, A.M.: Simulations of Dynamic Sequential Search Algorithms; C.ACM, 21(9):790-791, (Sep 1978). (3.1.3)

[621] Thanh, M. and Bui, T.D.: An Improvement of the Binary Merge Algorithm; BIT, 22(4):454-462, (1982). (4.3.3)

[622] Ting, T.C. and Wang, Y.W.: Multiway Replacement Selection Sort with Dynamic Reservoir; Computer Journal, 20(4):298-301, (Nov 1977). (4.4.0.2)

[623] Todd, S.: Algorithm and Hardware for a Merge Sort Using Multiple Processors; IBM J Res. Development, 22(5):509-517, (Sep 1978). (4.2.1)

[624] Trabb Pardo, L.: Stable Sorting and Merging with Optimal Space and Time Bounds; SIAM J on Computing, 6(2):351-372, (June 1977). (4.3.2, 4.1)

[625] Tropf, H. and Herzog, H.: Multidimensional Range Search in Dynamically Balanced Trees; Angewandte Informatik, 2:71-77, (1981). (3.5)

[626] Ullman, J.D.: A Note on the Efficiency of Hashing Functions; J.ACM, 19(3):569-575, (July 1972). (3.3)

[627] Ulrich, E.G.: Event Manipulation for Discrete Simulations Requiring Large Number of Events; C.ACM, 21(9):777-785, (Sep 1978); also 23(3):180-181, (Mar 1980). (5.1.9)

[628] Unterauer, K.: Dynamic Weighted Binary Search Trees; Acta Informatica, 11(4):341-362,

(1979). (3.4.1.4)

[629] Vaishnavi, V.K., Kriegel, H.P. and Wood, D.: Height Balanced 2-3 Trees; Computing, 21:195-211, (1979). (3.4.2.1)

[630] Vaishnavi, V.K., Kriegel, H.P. and Wood, D.: Optimum Multiway Search Trees; Acta Informatica, 14(2):119-133, (1980). (3.4.6)

[631] vanEmde-Boas, P., Kaas, R. and Zijlstra, E.: Design and Implementation of an Efficient Priority Queue; Mathematical Systems Theory, 10:99-127, (1977). (5.1.4)

[632] vanEmde-Boas, P.: Preserving Order in a Forest in Less than Logarithmic Time and Linear Space; Inf. Proc. Letters, 6(3):80-82, (June 1977). (5.1.4)

[633] vanEmde-Boas, P.: Preserving Order in a Forest in Less than Logarithmic Time; Proceedings FOCS, Berkeley CA, 16:75-84, (Oct 1975). (5.1.4)

[634] vanEmden, M.H.: Algorithm 402, qsort; C.ACM, 13(11):693-694, (Nov 1970). (4.1.3)

[635] vanEmden, M.H.: Increasing the Efficiency of Quicksort; C.ACM, 13(9):563-567, (Sep 1970). (4.1.3)

[636] vanLeeuwen, J. and Overmars, M.H.: Stratified Balanced Search Trees; Acta Informatica, 18(4):345-359, (1983). (3.4.1, 3.4.2)

[637] vanLeeuwen, J. and Wood, D.: Dynamization of Decomposable Searching Problems; Inf. Proc. Letters, 10(2):51-56, (Mar 1980). (2.2, 3)

[638] Vitter, J.S.: A Shared-Memory Scheme for Coalesced Hashing; Inf. Proc. Letters, 13(2):77-79, (Nov 1981). (3.3.12)

[639] Vitter, J.S.: Analysis of Coalesced Hashing; PhD Dissertation, Stanford University, (Aug 1980). (3.3.12)

[640] Vitter, J.S.: Analysis of the Search Performance of Coalesced Hashing; J.ACM, 30(2):231-258, (Apr 1983). (3.3.12)

[641] Vitter, J.S.: Deletion Algorithms for Hashing that Preserve Randomness; Proceedings FOCS, Nashville TN, 22:127-132, (Oct 1981) also J. of Algorithms, 3(3):261-275, (Sep 1982). (3.3.12)

[642] Vitter, J.S.: Implementations for Coalesced Hashing; C.ACM, 25(12):911-926, (Dec 1982). (3.3.12)

[643] Vitter, J.S.: Tuning the Coalesced Hashing Method to Obtain Optimum Performance; Proceedings FOCS, Syracuse NY, 21:238-247, (Oct 1980). (3.3.12)

[644] Vuillemin, J.: A Data Structure for Manipulating Priority Queues; C.ACM, 21(4):309-314, (Apr 1978). (5.1.7)

[645] Vuillemin, J.: A Unifying Look at Data Structures; C.ACM, 23(4):229-239, (Apr 1980). (2.1)

[646] Walker, A. and Wood, D.: Locally Balanced Binary Trees; Computer Journal, 19(4):322-325, (Nov 1976). (3.4.1.6)

[647] Warren, H.S.: Minimal Comparison Sorting by Choosing Most Efficient Comparisons; Inf. Proc. Letters, 2(5):129-130, (Dec 1973). (4.1.8)

[648] Waters, S.J.: Analysis of Self-Indexing, Disc Files; Computer Journal, 18(3):200-205, (Aug 1975). (3.2.2)

[649] Wegman, M.N. and Carter, J.L.: New Classes and Applications of Hash Functions; Proceedings FOCS, San Juan PR, 20:175-182, (Oct 1979). (3.3)

[650] Wegner, L.M: Sorting a Linked List with Equal Keys; Inf. Proc. Letters, 15(5):205-208, (Dec 1982). (4.2.2)

[651] Wessner, R.L.: Optimal Alphabetic Search Trees with Restricted Maximal Height; Inf. Proc. Letters, 4(4):90-94, (Jan 1976). (3.4.1.7)

[652] Whitt, J.D. and Sullenberger, A.G.: The Algorithm Sequential Access Method: an Alternative to Index Sequential; C.ACM, 18(3):174-176, (Mar 1975). (3.2.2, 3.4.3)

[653] Wiele, J.P. van de: An Optimal Lower Bound on the Number of Total Operations to Compute 0-1 Polynomials Over the Field of Complex Numbers; Proceedings FOCS, Ann Arbor MI, 19:159-165, (Oct 1978). (6.4)

[654] Wikstrom, A.: Optimal Search Trees and Length Restricted Codes; BIT, 19(4):518-524, (1979). (3.4.1.7)

[655] Willard, D.E.: Maintaining Dense Sequential Files in a Dynamic Environment; Proceedings SIGACT, San Francisco CA, 14:114-121, (May 1982). (3.1.1)

[656] Willard, D.E.: Polygon Retrieval; SIAM J on Computing, 11(1):149-165, (Feb 1982). (3.5)

[657] Williams, F.A.: Handling Identifiers as Internal Symbols in Language Processors; C.ACM, 2(6):21-24, (June 1959). (3.3.12)

[658] Williams, J.W.J.: Algorithm 232; C.ACM, 7(6):347-348, (June 1964). (4.1.5, 5.1.3)

[659] Wilson, L.B.: Sequence Search Trees: Their Analysis Using Recurrence Relations; BIT, 16(3):332-337, (1976). (3.4.1.1, 3.4.6)

[660] Winograd, S.: A New Algorithm for Inner Product; IEEE Trans. on Computers, C-17(7):693-694, (July 1968). (6.3)

[661] Winograd, S.: The Effect of the Field of Constants on the Number of Multiplications; Proceedings FOCS, Berkeley CA, 16:1-2, (Oct 1975). (6.2)

[662] Wise, D.S.: Referencing Lists by an Edge; C.ACM, 19(6):338-342, (June 1976). (3.1.1)

[663] Wong, J.K.: Some Simple In-Place Merging Algorithms; BIT, 21(2):157-166, (1981). (4.3.2)

[664] Woodall, A.D.: A Recursive Tree Sort; Computer Journal, 14(1):103-104, (1971). (4.2.6)

[665] Wright, W.E.: Binary Search Trees in Secondary Memory; Acta Informatica, 15(1):3-17, (1981). (3.4.1.1, 3.4.1.3)

[666] Yao, A.C-C. and Yao, F.F.: Lower Bounds on Merging Networks; J.ACM, 23(3):566-571, (July 1976). (4.3)

[667] Yao, A.C-C. and Yao, F.F.: On the Average-Case Complexity of Selecting k-th Best; Proceedings FOCS, Ann Arbor MI, 19:280-289, (Oct 1978); also SIAM J on Computing, 11(3):428-447, (Aug 1982). (5.2)

[668] Yao, A.C-C. and Yao, F.F.: The Complexity of Searching an Ordered Random Table;

Proceedings FOCS, Houston TX, 17:173-177, (Oct 1976). (3.2.2)

[669] Yao, A.C-C.: A Note on the Analysis of Extendible Hashing; Inf. Proc. Letters, 11(2):84-86, (Oct 1980). (3.3.13)

[670] Yao, A.C-C.: An Analysis of (h,k, 1)-Shellsort; J. of Algorithms, 1(1):14-50, (1980). (4.1.4)

[671] Yao, A.C-C.: On Random 2-3 Trees; Acta Informatica, 9(2):159-170, (1978). (3.4.2.1)

[672] Yao, A.C-C.: On the Evaluation of Powers; SIAM J on Computing, 5(1):100-103, (Mar 1976). (6.2)

[673] Yao, A.C-C.: Probabilistic Computations - Toward a Unified Measure of Complexity; Proceedings FOCS, Providence RI, 18:222-226, (Oct 1977). (2.2.2.1)

[674] Yao, A.C-C.: Should Tables Be Sorted?; Proceedings FOCS, Ann Arbor MI, 19:22-27, (Oct 1978); also J.ACM, 28(3):615-628, (July 1981). (3.2.1)

[675] Yao, A.C-C.: Space-Time Tradeoff for Answering Range Queries; Proceedings SIGACT, San Francisco CA, 14:128-136, (May 1982). (3)

[676] Yap, C.K.: New Upper Bounds for Selection; C.ACM, 19(9):501-508, (Sep 1976). (5.2)

[677] Yongjin, Z. and Jianfang, W.: On Alphabetic-Extended Binary Trees with Restricted Path Length; Scientia Sinica, 22(12):1362-1371, (Dec 1979). (3.4.1)

[678] Yuval, G.: A Simple Proof of Strassen's Result; Inf. Proc. Letters, 7(6):285-286, (Oct 1978). (6.3)

[679] Zave, D.A.: Optimal Polyphase Sorting; SIAM J on Computing, 6(1):1-39, (Mar 1977). (4.4.3)

[680] Ziviani, N. and Tompa, F.W.: A Look at Symmetric Binary B-trees; Infor, 20(2):65-81, (May 1982). (3.4.1.3, 3.4.2.2)

[681] Zobrist, A.L. and Carlson, F.R.: Detection of Combined Occurrences; C.ACM, 20(1):31-35, (Jan 1977); also 20(9):678-680, (Sep 1977). (3)

[682] Zvegintzov, N.: Partial-Match Retrieval in an Index Sequential Directory; Computer Journal, 23(1):37-40, (Feb 1980). (3.4.3)

[683] Zweben, S.H. and McDonald, M.A.: An Optimal Method for Deletions in One-Sided Height-Balanced Trees; C.ACM, 21(6):441-445, (June 1978). (3.4.1.3)

Appendix IV: ALGORITHMS CODED IN PASCAL AND IN C

The following entries are selected algorithms which are coded in a language different from that used in the main entries.

Binary search for arrays

```
for ( low=0, high=n;  high-low > 1;  )
        {
        i = (high+low) / 2;
        if ( key ≤ r[i].k )  high = i;
               else           low  = i;
        }
if ( key==r[high].k )      found( r[high] );
        else             notfound( key );
```

Insertion in a sorted array

```
        if ( n≥m )   Error /*** Table is full ***/;
        else  { for ( i=n++;  i>0 && r[i].k>key;  i-- )      r[i+1] = r[i];
                r[i+1].k = key;
                }
```

Linear probing hashing: search

```
        i = hashfunction( key ) ;
        last = (i+n-1) % m;
        while ( i≠last && !empty(r[i]) && r[i].k≠key )
                i = (i+1) % m;
        if (r[i].k==key)   found(r[i]);
                else    notfound(key);
```

Linear probing hashing: insertion

```
i = hashfunction( key ) ;
last = (i+m-1) % m;
while ( i≠last && !empty(r[i]) && !deleted(r[i]) && r[i].k≠key )
        i = (i+1) % m;
if (empty(r[i]) || deleted(r[i]))
                {
                /*** insert here ***/
                r[i].k = key;
                n++;
                }
else    Error /*** table full, or key already in table ***/;
```

Double hashing: search

```
i = hashfunction( key ) ;
inc = increment( key );
last = (i+(n-1)*inc) % m;
while ( i≠last && !empty(r[i]) && r[i].k≠key )
        i = (i+inc) % m;
if (r[i].k==key)    found(r[i]);
        else    notfound(key);
```

Double hashing: insertion

```
i = hashfunction( key ) ;
inc = increment( key );
last = (i+(m-1)*inc) % m;
while ( i≠last && !empty(r[i]) && !deleted(r[i]) && r[i].k≠key )
        i = (i+inc) % m;
if (empty(r[i]) || deleted(r[i]))
                {
                /*** insert here ***/
                r[i].k = key;
                n++;
                }
```

```
    else    Error /*** table full, or key already in table ***/;
```

Brent's reorganization scheme: insertion

```
init = hashfunction( key );
inc  = increment( key );
for (i=0; i≤n; i++)
        for (j=i; j≥0; j--)
                {
                jj = (init + j*inc) % m;
                ii = (jj + (i-j)*increment(r[jj].k)) % m;
                if ( empty(r[ii]) || deleted(r[ii]) )
                        {
                        /*** move record forward ***/
                        r[ii] = r[jj];
                        /*** insert new in r[jj] ***/
                        r[jj].k = key;
                        n++;
                        return;
                        }
                };
/*** table is full ***/
Error;
```

Data structure definition for Binary trees (C):

```
typedef struct btnode {               /*** Binary Tree Definition ***/
        typekey k;            /*** Key ***/
        struct btnode *left,*right; /*** Pointers to subtrees ***/
        } node, *tree;
```

Binary tree search (C):

```
search( key, t )
typekey key;
tree t;
{
while( t ≠ NULL )
        if ( t→k == key )
                { found( t ); return; }
        else if ( t→k < key )        t = t→right;
                        else    t = t→left;
notfound( key );
};
```

Binary tree insertion (C):

```
tree   insert( key, t )
typekey key;
tree t;
{
if ( t==NULL )     t = CreateNode( key, NULL, NULL );
else if ( t→k == key )
        Error; /*** Key already in table ***/
else if ( t→k < key )        t→right = insert( key, t→right );
                else    t→left  = insert( key, t→left );
return( t );
};
```

Note that the insertion algorithm returns the new tree, as "C" does not have **var** variables.

The Pascal data structure used to define B-trees is

B-tree data structure (Pascal)

```
btree =         ↑node;
node  =         record
        d : 0..2*M;
        k : array [1..2*M] of typekey;
        p : array [0..2*M] of btree
        end;
```

Note that the lexicographical order is given by the fact that all the keys in the subtree pointed by p[i] are greater than k[i] and less than k[i+1].

B-tree search (Pascal):

```
procedure search( key : typekey; t : btree );

var i : integer;
begin
if t=nil then {*** Not Found ***}
              notfound( key )
else    with t↑ do begin
        i := 1;
        while ( i<d ) and ( key>k[i] ) do        i := i+1;
        if key = k[i] then    {*** Found ***}
                              found( t↑, i )
        else    if key < k[i] then   search( key, p[i-1] )
                           else      search( key, p[i] )
        end
end;
```

B-tree insertion (Pascal):

```
function CreateNode( k1 : typekey; p0, p1 : btree ) : btree;

var     t : btree;
begin
        new( t );
        t↑.p[0] := p0;
        t↑.p[1] := p1;
        t↑.k[1] := k1;
        t↑.d := 1;
        CreateNode := t
end;

procedure insert( key : typekey; var t : btree );

var     ins : typekey;
        NewNode : btree;

  function InternalInsert( t : btree ) : typekey;
```

```
   var  i, j : integer;
        ins : typekey;
        tempr : btree;
   begin
   if t=nil then begin{*** The bottom of the tree has been reached:
                           indicate insertion to be done ***}
              InternalInsert := key;
              NewNode := nil
              end
   else with t↑ do begin
        InternalInsert := NoKey;
        i := 1;
        while ( i<d ) and ( key>k[i] ) do       i := i+1;
        if key = k[i] then
              Error {*** Key already in table ***}
        else  begin
              if key > k[i] then i := i+1;
              ins := InternalInsert( p[i-1] );
              if ins ≠ NoKey then
              {*** the key in "ins" has to be inserted in present node ***}
                   if d<2*M then      InsInNode( t, ins, NewNode )
                   else    {*** Present node has to be split ***}
                           begin
                           {*** Create new node ***}
                           if i≤M+1 then begin
                                tempr := CreateNode( k[2*M], nil, p[2*M] );
                                d := d-1;
                                InsInNode( t, ins, NewNode )
                                end
                           else tempr := CreateNode( ins, nil, NewNode );
                           {*** move keys and pointers ***}
                           for j:=M+2 to 2*M do
                                InsInNode( tempr, k[j], p[j] );
                           d := M;
                           tempr↑.p[0] := p[M+1];
                           InternalInsert := k[M+1];
                           NewNode := tempr
                           end
              end
        end
   end;

begin
     ins := InternalInsert( t );
     {*** check for growth at the root ***}
     if ins ≠ NoKey then t := CreateNode( ins, t, NewNode)
end;
```

The insertion code uses the function "InsertInNode", described below.

Insert an entry in a B-tree node (Pascal):

```
procedure InsInNode( t : btree; key : typekey; ptr : btree );

label  999;
var    j : integer;
begin
with t↑ do begin
       j := d;
       while j ≥ 1 do
              if key < k[j] then begin
                     k[j+1] := k[j];
                     p[j+1] := p[j];
                     j := j-1
                     end
              else   goto 999;  {*** break ***}
       999:
       k[j+1] := key;
       p[j+1] := ptr;
       d := d+1
       end
end;
```

Auxiliary functions for B-tree insertion (C)

```
btree CreateNode( k1, p0, p1 )
typekey k1;
btree p0, p1;

{btree tempr;
       tempr = (btree)malloc( sizeof( node ) );
       tempr->p[0] = p0;
       tempr->p[1] = p1;
       tempr->k[0] = k1;
       tempr->d = 1;
       return( tempr );
}
```

```
InsInNode( t, key, ptr )
btree t, ptr;
typekey key;

{int j;
        for( j=t->d; j>0 && key<t->k[j-1]; j-- )    {
                t->k[j] = t->k[j-1];
                t->p[j+1] = t->p[j];
                }
        t->d++;
        t->k[j] = key;
        t->p[j+1] = ptr;
}
```

4.1.2: Linear insertion sort (Pascal):

```
procedure sort( var r : ArrayToSort; lo, up : integer );

var     i, j : integer;
        tempr : ArrayEntry;
        flag : boolean;
begin
for i:=up-1 downto lo do begin
        tempr := r[i];
        j := i+1;
        flag := true;
        while (j≤up) and flag do
                if tempr.k > r[j].k then begin
                        r[j-1] := r[j];
                        j := j+1
                        end
                else    flag := false;
        r[j-1] := tempr
        end
end;
```

The above algorithm is slightly more complicated than the C version, as the internal loop cannot test for the double condition in a single statement. This forces the use of the boolean variable *flag*.

4.1.2: Linear insertion sort with sentinel (Pascal):

```
procedure sort( var r : ArrayToSort; lo, up : integer );

var     i, j : integer;
        tempr : ArrayEntry;
begin
r[up+1].k := MaximumKey;
for i:=up-1 downto lo do begin
        tempr := r[i];
        j := i+1;
        while tempr.k > r[j].k do begin
                r[j-1] := r[j];
                j := j+1
                end;
        r[j-1] := tempr
        end
end;
```

4.1.3: Quicksort (with bounded stack usage)

```
procedure sort( var r : ArrayToSort; lo, up : integer );

var     i, j : integer;
        tempr : ArrayEntry;
begin
while up>lo do begin
        i := lo;
        j := up;
        tempr := r[lo];
        {*** Split file in two ***}
        while i<j do begin
                while r[j].k > tempr.k do
                        j := j-1;
                r[i] := r[j];
                while (i<j) and (r[i].k<=tempr.k) do
                        i := i+1;
                r[j] := r[i]
                end;
        r[i] := tempr;
        {*** Sort recursively, the smallest first ***}
        if i-lo < up-i then begin
                        sort(r,lo,i-1);
```

```
                        lo := i+1
                        end
                else    begin
                        sort(r,i+1,up);
                        up := i-1
                        end
        end
end;
```

The above version of Quicksort is designed to prevent the growth of the recursion stack in the worst case (which could be $O(n)$). This is achieved by changing the second recursive call into a while-loop, and selecting the smallest array to be sorted recursively.

4.1.4: Shellsort (Pascal):

```
procedure sort( var r : ArrayToSort; lo, up : integer );

label   999;
var     d, i, j : integer;
        tempr : ArrayEntry;
begin
d := up-lo+1;
while d>1 do begin
        if d<5 then  d := 1
                else   d := trunc( 0.45454*d );
        {*** Do linear insertion sort in steps size d ***}
        for i:=up-d downto lo do begin
                tempr := r[i];
                j := i+d;
                while j ≤ up do
                        if tempr.k > r[j].k then begin
                                r[j-d] := r[j];
                                j := j+d
                                end
                        else    goto 999;  {*** break ***}
                999:
                r[j-d] := tempr
                end
        end
end;
```

As this algorithm is a composition using Linear insertion sort, the same comments can be applied.

For a predetermined, not computable, sequence of increments, the Shellsort algorithm becomes:

Shellsort for fixed increments

```
int Increments[] = {34807,15823,7193,3271,1489,677,307,137,61,29,13,5,2,1,0};
	sort( r, lo, up )
	ArrayToSort r;
	int lo, up;

	{int d, i, id, j;
	ArrayEntry tempr;
	for ( id=0; (d=Increments[id]) > 0; id++ ) {
		/*** Do linear insertion sort in steps size d ***/
		for ( i=up-d; i≥lo; i-- ) {
			tempr = r[i];
			for ( j=i+d; j≤up && (tempr.k>r[j].k); j+=d )
				r[j-d] = r[j];
			r[j-d] = tempr;
			}
		}
	}
```

Heapsort (C)

```
	sort( r, lo, up )
	ArrayToSort r;
	int lo, up;

	{int i;
	/*** construct heap ***/
	for ( i=up/2; i>1; i-- )  siftup(r,i,up);
	/*** repeatedly extract maximum ***/
	for ( i=up; i>1; i-- )    {
		siftup(r,1,i);
		exchange( r, 1, i );
		}
	};
```

Linear probing sort

```
sort( r, lo, up )
ArrayToSort r;
int lo, up;

{ArrayToSort r1;
int i, j, uppr;
uppr = up + (UppBoundr-up)*3/4;
for (j=lo; j<=up; j++)      r1[j] = r[j];
for (j=lo; j<=UppBoundr; j++)   r[j].k = NoKey;
for (j=lo; j<=up; j++) {
        for ( i=phi(r1[j].k,lo,uppr); r[i].k != NoKey; i++) {
                if ( r1[j].k < r[i].k ) {
                        r1[j-1] = r[i];
                        r[i] = r1[j];
                        r1[j] = r1[j-1];
                        };
                if ( i > UppBoundr ) Error;
                }
        r[i] = r1[j];
        };
for (j=i=lo; j<=UppBoundr; j++)
        if ( r[j].k != NoKey )
                r[i++] = r[j];
while ( i <= UppBoundr )
                r[i++].k = NoKey;
};
```

Merge sort (Pascal):

```
function sort( var r : list; n : integer ) : list;
label 999;
var    fi, la, temp : list;

begin
if r = nil then sort := nil
else    if n>2 then
        sort := merge( sort( r, n div 2 ),
                       sort( r, (n+1) div 2 ))
else    begin
        fi := r;        la := r;
```

```
        r := r↑.next;
        {*** Build list as long as possible ***}
        while r ≠ nil do
                if r↑.k ≥ la↑.k then begin
                        la↑.next := r;
                        la := r;
                        r := r↑.next;
                        end
                else if r↑.k ≤ fi↑.k then begin
                        temp := r;
                        r := r↑.next;
                        temp↑.next := fi;
                        fi := temp
                        end
                else    goto 999;
999:
        la↑.next := nil;
        sort := fi
        end
end;
```

The above algorithm is similar to the one in the main entry, except that at the bottom level of recursion, it tries to construct the longest possible list of ordered elements. To achieve this, it compares the next element in the list against the head and the tail of the list being constructed. Consequently, this algorithm will improve significantly when used to sort partially ordered (or reverse-ordered) files.

Merge sort (C):

```
list sort( n )
int n;

{
list fi, la, temp;
extern list r;
if ( r == NULL ) return( NULL );
else    if ( n>1 )
        return( merge( sort( n/2 ), sort( (n+1)/2 )));
else    {
        fi = r;la = r;
        /*** Build list as long as possible ***/
        for ( r=r→next; r≠NULL; )
                if ( r→k ≥ la→k ) {
                        la→next = r;
```

```
                        la = r;
                        r = r→next;
                        }
                else if ( r→k ≤ fi→k ) {
                        temp = r;
                        r = r→next;
                        temp→next = fi;
                        fi = temp;
                        }
                else    break;
        la→next = NULL;
        return( fi );
        }
};
```

Due to the absence of **var** variables in C, the list to be sorted is stored in a global variable named *r*.

Radix sort (C):

```
list sort( r )
list r;

{
list head[M], tail[M];
int i, j, h;
for (i=D; i>0; i--) {
        for (j=0; j<M; j++)     head[j] = NULL;
        while ( r ≠ NULL ) {
                h = charac( i, r→k );
                if ( head[h]==NULL ) head[h] = r;
                else    tail[h]→next = r;
                tail[h] = r;
                r = r→next;
                };
        /*** Concatenate lists ***/
        r = NULL;
        for (j=M-1; j≥0; j--)
                if ( head[j] ≠ NULL ) {
                        tail[j]→next = r;
                        r = head[j];
                        }
        };
return( r );
};
```

The above algorithm uses the function *charac* which returns the $i sup th$ character of the given key. The global constant “M“ gives the range of the alphabet (or characters). The constant or variable “D“ gives the number of characters used by the key.

Top-down radix sort (C):

```
list sort( s, j )
list s;
int j;

{
int i;
list head[M], t;
struct rec aux;
extern list Last;
if (s==NULL)      return(s);
if ( s→next == NULL )   {Last = s;   return(s);}
if ( j>D )   {
        for (Last=s; Last→next≠NULL; Last = Last→next);
        return( s );
        }
for (i=0; i<M; i++)     head[i] = NULL;
/* Place records in buckets */
while (s ≠ NULL)  {
        i = charac( j, s→k );
        t = s;
        s = s→next;
        t→next = head[i];
        head[i] = t;
        }
/* sort recursively */
t = &aux;
for (i=0; i<M; i++)     if (head[i]≠NULL) {
        t→next = sort( head[i], j+1 );
        t = Last;
        }
return( aux.next);
}
```

List merging (C):

```
list merge( a, b )
list a, b;

{
list temp;
struct rec aux;
temp = &aux;
while ( b ≠ NULL )
        if ( a == NULL )  { a = b; break; }
        else    if ( b→k > a→k )
                            { temp = temp→next = a; a = a→next; }
                else        { temp = temp→next = b; b = b→next; };
temp→next = a;
return( aux.next );
};
```

Array merging into same or third array

```
procedure merge ( a,b : RecordArray; var c : RecordArray; na,nb : integer );
{***
        Merges the arrays a and b into c (increasing order assumed)
                a or b may coincide with c ***}
begin
while (na≥1) or (nb≥1) do
        if na<1 then
                while nb>0 do begin
                        c[nb] := b[nb];
                        nb := nb-1
                        end {while}
        else if nb<1 then
                while na>0 do begin
                        c[na] := a[na];
                        na := na-1
                        end {while}
        else if a[na].k < b[nb].k then begin
                c[na+nb] := b[nb];
                nb := nb-1
                end {if...then}
        else begin
                c[na+nb] := a[na];
                na := na-1
```

```
          end; {else}
end;
```

5.1.1: Sorted list extraction

```
function extract( var pq : list ) : typekey;
begin
if pq=nil then Error {*** Extracting from empty queue ***}
else    begin
        extract := pq↑.k;
        pq := pq↑.next
        end
end;
```

5.1.1: Sorted list insertion

```
procedure insert( new : list; var pq : list );
label   9999;
var     p : list;

begin
if pq=nil then pq := new
else if pq↑.k < new↑.k then begin
        new↑.next := pq;
        pq := new
        end
else    begin
        p := pq;
        while p↑.next ≠ nil do begin
                if p↑.next↑.k < new↑.k then begin
                        new↑.next := p↑.next;
                        p↑.next := new;
                        goto 9999
                        end;
                p := p↑.next
                end;
        p↑.next := new
        end;
9999:
end;
```

5.1.1: Unsorted list extraction

```
function extract( var pq : list ) : typekey;
var     max, p : list;

begin
if pq=nil then Error {*** Extraction from an empty list ***}
else if pq↑.next = nil then begin
        extract := pq↑.k;   pq := nil      end
else begin
        max := pq;  p := pq;
        while p↑.next ≠ nil do begin
                if max↑.next↑.k < p↑.next↑.k then max := p;
                p := p↑.next
                end;
        if max↑.next↑.k < pq↑.k then begin
                extract := pq↑.k;  pq := pq↑.next   end
        else begin
                extract := max↑.next↑.k;
                max↑.next := max↑.next↑.next
                end
        end
end;
```

5.1.1: Unsorted list insertion

```
procedure insert( new : list; var pq : list );
begin
new↑.next := pq;
pq := new
end;
```

5.1.2: P-trees deletion

```
procedure delete ( var pq : tree );
begin
if pq = nil then Error {*** deletion on an empty queue ***}
else if pq↑.left = nil then pq := nil
else if pq↑.left↑.left = nil then begin
        pq↑.left := pq↑.right;
        pq↑.right := nil
        end
else    delete( pq↑.left )
end;
```

5.1.2: P-trees insertion

```
procedure insert ( new : tree; var pq : tree );
label 9999;
var   p : tree;

begin
if pq = nil then      pq := new
else if pq↑.k ≥ new↑.k then begin
                {*** Insert above subtree ***}
                new↑.left := pq;
                pq := new
                end
else    begin
        p := pq;
        while p↑.left ≠ nil do
                if p↑.left↑.k ≥ new↑.k then begin
                        {*** Insert in right subtree ***}
                        insert( new, p↑.right );
                        goto 9999
                        end
                else    p := p↑.left;
        {*** Insert at bottom left ***}
        p↑.left := new
        end;
9999:
end;
```

5.1.2: P-trees, inspection of top of queue

```
function inspect ( pq : tree ) : typekey;
begin
if pq = nil then Error {*** Inspecting an empty queue ***};
while pq↑.left ≠ nil do pq := pq↑.left;
inspect := pq↑.k
end;
```

5.1.3: Heap insertion

```
insert( new, r )
RecordArray r;
ArrayEntry new;

{int i, j;
extern int n;
n++;
for (j=n; j>1; j=i) {
        i = j/2;
        if ( r[i].k ≥ new.k ) break;
        r[j] = r[i];
        }
r[j] = new;
};
siftup( r, i, n )
RecordArray r;
int i, n;

{ArrayEntry tempr;
int j;
        while ( (j=2*i) ≤ n ) {
                if ( j<n && r[j].k < r[j+1].k ) j++;
                if ( r[i].k < r[j].k ) {
                        tempr = r[j];
                        r[j] = r[i];
                        r[i] = tempr;
                        i = j;
                        }
                else    break;
                }
};
```

5.1.3: Heap deletion

```
delete( r )
RecordArray r;

{
extern int n;
if ( n<1 ) Error /*** extracting from an empty Heap ***/;
else    {
        r[1] = r[n];
        siftup( r, 1, --n );
        }
};
```

5.1.5: Pagodas insertion

```
tree merge( a, b )
tree a, b;

{
tree bota, botb, r, temp;
if ( a==NULL ) return( b );
else if ( b==NULL ) return( a );
else    {
        /*** Find bottom of a's rightmost edge ***/
        bota = a->right;    a->right = NULL;
        /*** bottom of b's leftmost edge ***/
        botb = b->left;     b->left = NULL;
        r = NULL;
        /*** Merging loop ***/
        while ( bota!=NULL && botb!=NULL )
                if ( bota->k < botb->k ) {
                        temp = bota->right;
                        if ( r==NULL )    bota->right = bota;
                                else    {bota->right = r->right;
                                        r->right = bota;
                                        };
                        r = bota;
                        bota = temp;
```

```
                    }
            else    {temp = botb→left;
                    if ( r==NULL )    botb→left = botb;
                            else    {botb→left = r→left;
                                    r→left = botb;
                                    };
                    r = botb;
                    botb = temp;
                    };
        /*** one edge is exhausted, finish merge ***/
        if ( botb==NULL ) {
                a→right = r→right;
                r→right = bota;
                return( a );
                }
        else    {b→left = r→left;
                r→left = botb;
                return( b );
                }
        }
};

tree insert( new, pq )
tree new, pq;
{
new→left = new;    new→right = new;
return( merge( pq, new ) );
};
```

5.1.5: Pagodas deletion

```
tree delete( pq )
tree pq;

{
tree le, ri;
if ( pq==NULL ) Error /*** Deletion on empty queue ***/;
else    {
        /*** Find left descendant of root ***/
        if ( pq→left == pq ) le = NULL;
                else {
                        le = pq→left;
                        while ( le→left ≠ pq ) le = le→left;
                        le→left = pq→left;
```

```
                };
        /*** Find right descendant of root ***/
        if ( pq→right == pq ) ri = NULL;
              else {
                    ri = pq→right;
                    while ( ri→right ≠ pq ) ri = ri→right;
                    ri→right = pq→right;
                    };
        /*** merge them ***/
        return( merge( le, ri ) );
        }
};
```

5.1.6.1: Leftist trees deletion

```
tree merge( a, b )
tree a, b;
{
if ( a == NULL ) return( b );
else if ( b == NULL ) return( a );
else if ( a→k > b→k ) {
        a→right = merge( a→right, b );
        fixdist( a );
        return( a );
        }
else    {
        b→right = merge( a, b→right );
        fixdist( b );
        return( b );
        }
};

tree delete( pq )
tree pq;
{
if ( pq == NULL ) Error /*** delete on an empty queue ***/;
else    return( merge( pq→left, pq→right ) );
};
```

5.1.6.1: Leftist trees insertion

```
tree insert( new, pq )
tree new, pq;

{
if ( pq==NULL )   return( new );
else if ( pq→k > new→k ) {
            pq→right = insert( new, pq→right );
            fixdist( pq );
            return( pq );
            }
else    {
        new→left = pq;
        return( new );
        }
};
```

5.1.6.2: Binary priority queues deletion

```
tree delete( pq )
tree pq;

{tree temp;
if ( pq == NULL ) Error /*** deletion on an empty queue ***/;
else if ( pq→right == NULL )
        return( pq→left );
else    {
        /*** promote left descendant up ***/
        pq→k = pq→left→k;
        pq→left = delete( pq→left );
        /*** rearrange according to constraints ***/
        if ( pq→left == NULL ) {
                pq→left = pq→right;     pq→right = NULL; };
        if ( pq→right ≠ NULL )
                if ( pq→left→k < pq→right→k ) {
                        /*** descendants in wrong order ***/
                        temp = pq→right;
                        pq→right = pq→left;
                        pq→left = temp;
                        }
        return( pq );
        }
};
```

5.1.6.2: Binary priority queues insertion

```
tree insert( new, pq )
tree new, pq;

{
if ( pq == NULL ) return( new );
else if ( pq→k ≤ new→k ) {
                new→left = pq;
                return( new );
                }
else if ( pq→left == NULL )
                pq→left = new;
else if ( pq→left→k ≤ new→k )
                pq→left = insert( new, pq→left );
else            pq→right = insert( new, pq→right );
return( pq );
};
```

5.1.6.2: Mergin of binary priority queues

```
function merge ( a, b : tree ) : tree;
var     temp : tree;
begin
if a=nil then merge := b
else if b=nil then merge := a
else    begin
        if a↑.k < b↑.k then begin
                temp := a;   a := b;   b := temp  end;
        a↑.right := merge( a↑.right, b );
        if a↑.left ≠ nil then
                if a↑.left↑.k < a↑.right↑.k then begin
                        temp := a↑.right;
                        a↑.right := a↑.left;
                        a↑.left := temp
                        end
        end
end;
```

Index